Studies in Computational Intelligence

Volume 632

Series editor

Janusz Kacprzyk, Polish Academy of Sciences, Warsaw, Poland
e-mail: kacprzyk@ibspan.waw.pl

About this Series

The series "Studies in Computational Intelligence" (SCI) publishes new developments and advances in the various areas of computational intelligence—quickly and with a high quality. The intent is to cover the theory, applications, and design methods of computational intelligence, as embedded in the fields of engineering, computer science, physics and life sciences, as well as the methodologies behind them. The series contains monographs, lecture notes and edited volumes in computational intelligence spanning the areas of neural networks, connectionist systems, genetic algorithms, evolutionary computation, artificial intelligence, cellular automata, self-organizing systems, soft computing, fuzzy systems, and hybrid intelligent systems. Of particular value to both the contributors and the readership are the short publication timeframe and the worldwide distribution, which enable both wide and rapid dissemination of research output.

More information about this series at http://www.springer.com/series/7092

Roman Gumzej

Engineering Safe and Secure Cyber-Physical Systems

The Specification PEARL Approach

Roman Gumzej
Faculty of Logistics
University of Maribor
Celje
Slovenia

ISSN 1860-949X   ISSN 1860-9503   (electronic)
Studies in Computational Intelligence
ISBN 978-3-319-28903-8   ISBN 978-3-319-28905-2   (eBook)
DOI 10.1007/978-3-319-28905-2

Library of Congress Control Number: 2015960778

Printed on acid-free paper

This Springer imprint is published by SpringerNature
The registered company is Springer International Publishing AG Switzerland

This book is dedicated to Eva

Foreword

The term cyber-physical system (CPS) is just a few years old and is now in vogue. It refers, however, by no means to anything new. Actually, it is a synonym for real-time computer system, which was defined by J. Martin already half a century ago as "one which controls an environment by receiving data, processing them, and taking action or returning results sufficiently quickly to affect the functioning of the environment at that time." Particularly with respect to the attribute "sufficiently quickly" this definition was refined by a German standard: "Real-time operation is the operating mode of a computing system, in which the programs for the processing of data arriving from the outside are permanently ready in such a way, that the processing results become available within time periods given a priori; these data may become available for processing either at randomly distributed instants or at predetermined points in time." In these definitions, the computer stands for the cyber-part of a CPS controlling its physical part, viz. the "environment" or the "outside", with the dynamics there the computer must keep pace.

To support the design of embedded real-time systems, a number of computer-aided tools were developed. They are, however, generally not satisfactory for different reasons. Some were derived from tools originally developed for the non-real-time domain by just adding to them an often insufficient minimum of real-time functionality. Others were too formal and, thus, not appealing to engineers for use in real-life projects. In this situation the book in hand presents as a remedy a novel approach based on the Process and Experiment Automation Real-time Language (PEARL). The development of this programming language commenced in 1969, starting out with a rather complex first design. Since then, a simplification process extensively exploiting experience gained in course of its industrial use led to several improved versions of PEARL.

A unique feature of PEARL is its closeness to natural language rendering PEARL code to be easily readable and understandable, even for persons who do not know the language. Furthermore, its very high-level constructs to a certain extent let it even be feasible for use as a specification language. Therefore, the author of this book bases his approach on this feature and derives from the programming language PEARL the language Specification PEARL. As graphical representations usually appeal better to humans than textual ones, the author makes his approach also applicable within the framework of the Unified Modeling Language (UML) exploiting the latter's extensibility. To this end, he maps PEARL's architectural constructs into UML and provides suitable stereotypes, profiles, and patterns.

Since real-time systems are very closely linked to processes in their environments and also physically located there, another synonymous term referring to them is embedded systems. Today, 98 % of all processors built are embedded in technical systems of any kind, where their task is to automate the operation of the environments embedding them. To a large extent, these automation functions are safety-related, or even safety-critical. That is also the reason why currently an initiative is under way to elaborate a novel version of PEARL, which will be the first programming language explicitly oriented at functional safety. Enclosed in its present version, in a nested way it will comprise four more and more restrictive partial languages, each one corresponding to one of the safety integrity levels as defined in an international safety standard.

But safety is not the only problem resulting from the very nature of automation systems, which must be taken into consideration in the course of their design. In contrast to earlier times, when embedded systems were proprietary and operating independently on their own, now they tend to be interconnected, often via the Internet. As a result, they are confronted with the same security problems as computers in the non-real-time domain, viz. malware intrusion and eavesdropping. The current trend toward complete interconnectedness as championed by initiatives such as Internet of Things or Industry 4.0 will even exacerbate the dangers to the informational security of embedded automation systems.

With this in mind, the book in hand presents a holistic approach oriented at quality of service and stressing the requirements of safety and security, in addition to the ones of correctness and timeliness, right from the start, i.e., immediately by design. In other words, the old path is left of trying to make already designed and implemented systems safe and secure, and to verify their correct and timely behavior at later stages. Instead, both by its notation and by its co-simulation features the Specification PEARL co-design methodology provides for self-documentation as well as verification and validation. In correspondence with the safety orientation of PEARL's forthcoming version, this methodology comprises guidelines for the appropriate use and parameterization of its constructs aiming to comply with the individual safety and security levels as defined by the standards pertaining for safety and security of cyber-physical systems.

Hagen   Wolfgang A. Halang
January 2016

Preface

Cyber-physical systems representing networked computational systems controlling physical entities build on the concepts of embedded and autonomous systems that can be enhanced by methods of artificial intelligence. They are spatially and temporally determined and need to be aware of that during their operation, for the signals from their environment to be adequately captured and assessed. They need to expose properties, native to autonomous systems: self-management, self-configuration, self-optimization, self-protection and self-healing. An important emphasis while using these systems lies with the concepts of their timeliness, functional correctness, safety of their operation as well as security of their transferred and stored data, which need to be assured according to appropriate standards on all levels of their operation. Hence, they need to be designed holistically by using the systems approach and engineering with respect to these standards.

The dependability of cyber-physical systems is usually assured by redundancy and over-scaled components. This results in more complex designs and higher costs, but often without guaranteeing safety or security. To achieve better overall quality, much effort was invested in the search for standardized components, methods and tools apt to improve the designed system's predictability and dependability. The design and development procedures of contemporary cyber-physical systems are well established, relatively cheap and widely used. Hardware components come with specifications, which undoubtedly state their capabilities and performance indicators. Complexity increases, however, when there is a need for their integration into larger set-ups and system-level performance must be assured. Software makes things even more complicated, as the WORE (Write-Once-Run-Everywhere) principle is hard to achieve, and different software engineering techniques can lead to programs with very different quality-of-service while running on the same hardware platform. To achieve a managed level of quality (of service), systems engineering methods should enable hardware–software co-design as well as efficient system's design and subsequent prototype verification and validation before putting them to use.

Throughout this book, a holistic quality of service-oriented approach to design and development of cyber-physical systems, with emphasis on their (timely) predictable and dependable behaviour, is presented and discussed. By following the standards for embedded system's safety and using appropriate hardware and software components inherently safe system's architectures can be devised and certified. At the same time their complexity is reduced to a reasonable level. The methodology and guidelines for designing and developing cyber-physical systems will result in their increased ability to be certified for safety and security as well as their improved interoperability.

Celje   Roman Gumzej
January 2015

Contents

Chapter 1
Introduction

1.1 Cyber-Physical Systems

Our society is facing considerable challenges in terms of climate change, energy efficiency, renewable energies, disease control, increasing traffic congestion, etc. Technology can play a major role in alleviating arising problems by the development of so-called smart infrastructures.

The idea behind smart infrastructures is to incorporate intelligence in everyday objects or services in order to improve the efficiency of performing certain rudimentary but crucial tasks. This trend of developing intelligent systems has already begun. A modern household incorporates more than 100 microprocessors (e.g. in vehicles, appliances, entertainment systems, cameras, wireless devices, personal digital assistants and toys), while a typical car alone includes already more than 50 microprocessors [1]. In fact, most microprocessors nowadays are embedded in systems not being computers [2]. The crucial technologies having made this leap possible are miniature sensing, communication and processing platforms, which can be embedded as parts in larger systems or processes to provide real-time monitoring and feedback control services [3]. Such platforms, deeply embedded in physical processes, are the so-called Cyber-Physical Systems (CPS).

CPS are being used in very different applications. Irrespective of the application domain, a CPS has three principal characteristics:

1. Environment Coupling: CPS are very tightly coupled with their environment (physical processes), i.e. any behavioural change in their environments results in a change in the CPS' behaviour and vice versa.
2. Diverse Capabilities: CPS are usually composed of diverse heterogeneous entities with capabilities differing by orders of magnitude. Sensors, deeply embedded in physical processes for monitoring purposes, have limited capabilities, while the entities managing them are much more capable. A direct consequence of this heterogeneity are potential bottlenecks in terms of computation, communication and memory capacity in CPS' workflows.

© Springer International Publishing Switzerland 2016
R. Gumzej, *Engineering Safe and Secure Cyber-Physical Systems*,
Studies in Computational Intelligence 632, DOI 10.1007/978-3-319-28905-2_1

3. Networked: Unlike traditional stand-alone embedded systems, CPS usually require communication channels between their components, either embedded within their physical processes or external to them, in order to provide their (usually coordinated) services [4].

Many Quality of Service (QoS) issues need to be addressed in order to make each of the before-mentioned CPS features viable, such as managing the cybernetical and physical processes and interactions, ensuring safety, energy efficiency, interoperability and sustainability. Two of the most important aspects to be considered at an early development phase of any CPS are its safety and security.

1.2 QoS of Cyber-Physical Systems

CPS operate in real-time and under real-time constraints. Hence, their QoS properties correspond with the QoS criteria for real-time systems. They have been systematically addressed in Gumzej [5]. Depending on the application, the rigour of compliance with individual QoS criteria may vary in type (quantitative or qualitative) and precision (low, high, daily, hourly or in milliseconds).

Considering the nature of contemporary CPS, we shall investigate their following QoS properties more closely

- correctness and timeliness,
- safety,
- security,
- ability to be licensable.

Correctness and timeliness are considered the most important characteristics of any CPS. They need to be assured in order to provide for CPS operation to be beneficial. Since synchronisation with their associated environment is a key issue in CPS design, the temporal predictability of their execution behaviour is considered as important for their overall correct operation as functional correctness. This can only be achieved by applying rigourous CASE methods enabling closed-loop system modelling, verification and validation to ensure their correct behaviour and temporal predictability.

In order to achieve the desired QoS properties, it was discovered early that a design methodology for CPS has to include the measures joined in the framework of the ISO/IEC 13236 [6] and related standards for QoS in information technology, which would ensure that the QoS criteria are considered during the CPS' entire life cycle. In the following sections appropriate safety and security assurance methods are described and the standards are listed, which may be applied to accordingly engineered systems.

1.2.1 Safety Integrity Levels

In the late 1980s, the International Electrotechnical Commission (IEC) started the standardisation of safety issues in computer control. Four *Safety Integrity Levels* (SIL1–SIL4) were defined, with SIL4 being the most critical one. Prescribed were activities at different levels and phases of system development (e.g. coding standards, dynamic analysis and testing, black-box testing, failure analysis, modelling, performance testing, formal methods, static analysis, modular approach), which are desired or mandatory, and approaches, which are allowed or required in order to fulfil the requirements of a certain Safety Integrity Level. These rules form the standard IEC 61508 for the life cycle management of instrumented protection systems. As can be seen from Fig. 1.1, the safety life cycle encompasses the entire production cycle from a system's design to its decommissioning.

The flowchart in Fig. 1.1 represents the safety life cycle of an Equipment Under Control (EUC) in its entirety. Such an EUC is composed of one or more Electrical/Electronic/Programmable Electronic (E/E/PE) devices, which have to fulfil individual as well as collective safety requirements as a system.

Apart from the above-mentioned process techniques to achieve system safety, some design techniques have also been devised. The latter, representing vital constituents of a system's development phase, form Parts 6 and 9 of the safety life cycle in Fig. 1.1. Some of them are summarised in Table 1.1 together with their importance to the individual Safety Integrity Levels.

In our case CPS and their constituent parts represent the EUC. Hence, the mentioned safety requirements with appropriate safety measures can be transferred to them. First we need to allocate the safety requirements in a CPS' life cycle. Then, we can apply the prescribed safety measures in a targeted manner to the CPS and its constituent parts. From the hardware architecture point of view they typically comprise, e.g. redundancy to ensure robustness. From the software architecture point of view they include some restrictions to the software design (c.p. Table 1.1), ensuring their dependability, and some fail-safe mechanisms for, e.g. error handling, graceful degradation, etc., ensuring their required predictability.

1.2.2 Security Capability Levels

As with safety, security measures also need to be incorporated into system CPS design by advance planning throughout their entire life cycle. They include risk assessment and security-related safety measures, like contingency planning, authentication or authorisation strategies, etc.

Being closely connected with their physical environments, CPS are very susceptible to targeted attacks. The systems controlled by CPS are very diverse in terms of scale and interconnectedness. The CPS can be as small as sensors (for, e.g., motion, temperature, pressure etc.) with lots of them scattered across an area, or as large

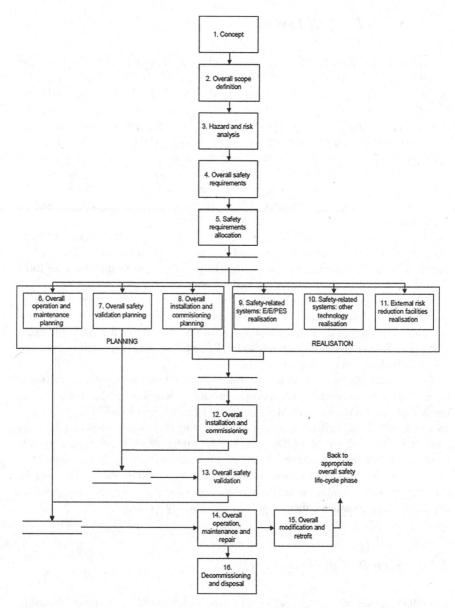

Fig. 1.1   Safety life cycle according to IEC 61508

as industrial control—process control (PCS) or supervision and control (SCADA)—systems (e.g. autonomously running power plants). Even a small scale security breach can result in severe consequences. As an example, let us consider pacemakers as CPS from an attacker's point of view. They are mainly meant to monitor the host person's

Table 1.1 Software practices from IEC 61508-3 by category

Practice	61508-3	SIL1	SIL2	SIL3	SIL4
Coding standards					
Use of coding standard	B.1	HR	HR	HR	HR
No dynamic variables	B.1	–	R	HR	HR
Dynamic analysis and testing					
Test case execution from cause consequence diagrams	B.2	–	–	R	R
Structure-based testing	B.2	R	R	HR	HR
Black-box testing					
Equivalence classes and input partition testing	B.3	R	HR	HR	HR
Failure analysis					
Failure modes, effects and criticality analysis	B.4	R	R	HR	HR
Formal methods modelling	B.5	–	R	R	HR
Performance modelling	B.5	R	HR	HR	HR
Timed Petri nets	B.5	–	R	HR	HR
Performance testing					
Avalanche/stress testing	B.6	R	R	HR	HR
Response timings and memory constraints	B.6	HR	HR	HR	HR
Performance requirements	B.6	HR	HR	HR	HR
Semi-formal methods					
Sequence diagrams	B.7	R	R	HR	HR
Finite state machines/state transition diagrams	B.7	R	R	HR	HR
Decision/truth tables	B.7	R	R	HR	HR
Static analysis					
Boundary value analysis	B.8	R	R	HR	HR
Control flow analysis	B.8	R	HR	HR	HR
Fagan inspections	B.8	–	R	R	HR
Symbolic execution	B.8	R	R	HR	HR
Walk-throughs/design reviews	B.8	HR	HR	HR	HR
Modular approach					
Software module size limit	B.9	HR	HR	HR	HR
Information hiding/encapsulation	B.9	R	HR	HR	HR
Fully defined interface	B.9	HR	HR	HR	HR
Total recommended (R)		12	12	3	1
Total highly recommended (HR)		6	10	20	22

Legend: HR highly recommended; R recommended; – no recommendation

heartbeat. Hence, they can be targeted to reveal a patient's electrocardiogram data in order to determine his or her physical condition. However, if tampered with, they can also be used to actuate an untimely shock that may harm the patient.

The interconnections between CPS may be local or limited to close proximities in case of wearable devices (such as the previously mentioned pacemaker) or sparse as in the example of sensor networks, electrical power grids or traffic control systems. They can be (hard) wired or wireless (radio). Hence, it is hard to generalise the potential impact of a CPS malfunction due to a security breach. They could affect only one person, an area or whole (groups of) countries.

From the examples indicated, we may conclude that CPS are often used to monitor and control mission critical processes. Therefore, any security compromise due to lacking protection of a CPS may have profound consequences for the system the CPS is associated with or embedded into. Moreover, since CPS have the ability to monitor the physical processes they control, this makes them privy to detailed and often sensitive information about the process. If this information becomes available to malicious entities, it can be exploited leading to loss of privacy and/or abuse and discrimination. Hence, the CPS' interfaces and communications need to be secured, so that only authorised persons and systems may access the process data.

In recent years, many organisations have collaborated to develop standards and guidelines on cyber-security for CPS. In 2002 the International Society of Automation (ISA) began writing a series of standards entitled ISA 99, which address the subject of cyber-security for industrial automation and control systems. Three standards have been released so far:

1. ANSI/ISA 99.01.01-2007 "Security for Industrial Automation and Control Systems Part 1: Terminology, Concepts, and Models" [7],
2. ANSI/ISA 99.02.01-2009 "Security for Industrial Automation and Control Systems: Establishing an Industrial Automation and Control Systems Security Program" [8], and
3. ANSI/ISA 99.03.03-2013 "Security for industrial automation and control systems Part 3-3: System security requirements and security levels" [9].

These standards describe the basic concepts and models related to cyber-security, as well as the elements contained in a cyber-security management system for use in the industrial automation and control systems environment. They also provide guidance on how to meet the requirements described for each element.

TC 65 WG 10 of the International Electrotechnical Commission (IEC) has joined with ISA 99 and will publish IEC versions of the standards under IEC 62443. Two have been published so far:

- IEC 62443-1-1 "Industrial communication networks—Network and system security - Part 1-1: Terminology, concepts and models" (related to ISA 99.01.01) [10] and
- IEC 62443-2-1 "Industrial communication networks—Network and system security - Part 2-1: Establishing an industrial automation and control system security program" [11] (related to ISA 99.02.01).

Over the next few years, these standards are expected to become the core standards for industrial control security worldwide.

In 2011 the "Guide to Industrial Control Systems (ICS) Security" was published by the National Institute of Science and Technology (NIST) as Special Publication 800-82 [12], and made available to general public. This document provides comprehensive industrial control system security guidance for various industries (electric, water and wastewater, oil and natural gas, chemical, pharmaceutical, pulp and paper, food and beverage, as well as discrete manufacturing, i.e. automotive, aerospace, and durable goods) and includes parts of the previously mentioned standards with guidelines for their application.

As with levels of safety (SIL, defined in the previous section), there are four system Security capability Levels (SL, defined by the before-mentioned standards):

SL1 Protection against causal or coincidental violation.
SL2 Protection against intentional violation using simple means.
SL3 Protection against intentional violation using sophisticated means.
SL4 Protection against intentional violation using sophisticated means with extended resources.

They provide measures that have to be in place on system and component levels to assure the corresponding security level (e.g. Table 1.2). They differ in rigour and extent of security measures implemented, resulting in different degrees of compliance with the SL.

In order to enable reasoning on CPS security, we need to consider their life cycles as well as their workflows. We have found out about their life cycles in the previous section. In general, CPS are autonomous systems [13] and, hence, we can categorise their workflows into four main functions:

1. *Monitoring*, being the most fundamental aspect of CPS, deals with sensing and gathering data from the environment in which a CPS is functioning; depending on the type of data and device, these data may also be (temporarily) stored in the CPS.

Table 1.2 Extract of ISA-99.03.03, draft 4

System requirement	SL
SR 1.1 The control system shall provide the capability to identify and authenticate all users (humans, software processes and devices). This capability shall enforce such identification and authentication on all interfaces which provide access to the control system to support segregation of duties and least privilege in accordance with applicable security policies and procedures	1
SR 1.1 RE 1 The control system shall provide the capability to *uniquely* identify and authenticate all users (humans, software processes and devices)	2
SR 1.1 RE 2 The control system shall provide the capability to employ multifactor authentication for human user access to the control system *via an untrusted network* (see 4.12, SR 1.10—Access via untrusted networks)	3
SR 1.1 RE 3 The control system shall provide the capability to employ multifactor authentication for *all* human user access to the control system	4

2. *Analysis* deals with analysing the data, collected during monitoring, to determine whether the physical process is meeting certain pre-defined criteria.
3. *Planning* is important in situations when the criteria are not satisfied; here, corrective actions are determined, which, when executed, ensure that the criteria are satisfied; it is also used to provide feedback on any past actions taken by the CPS, hereby enabling taking correct actions in the future; an underlying knowledge base may be used in order to determine the best actions.
4. *Execution* deals with the actuation of actions determined during the planning phase; it can take many forms from changing the cyber-behaviour of the CPS to controlling the physical process itself.

A CPS can operate in one of the three possible modes:

1. *Passive*: in this mode CPS act as information gathering platforms only, and solely monitor their environment, gather data and prepare them for processing.
2. *Semi-Active*: in this mode CPS monitor their environments (physical aspect) and analyse the data; if they detect some criteria not to be fulfilled, they execute indirect actions to change their own behaviour (cyber-aspect), so that the criteria can be satisfied.
3. *Active*: in this mode CPS monitor their physical environments and analyse the data; if they detect some criteria not to be fulfilled, they execute direct actions to modify the behaviour of the physical environments, so that the criteria are satisfied.

Given the recent trend towards complex and open designs, use of Commercial-Off-The-Shelf (COTS) components and interconnection via the existing insecure global communication infrastructure, such as the Internet, security has become very important for CPS. As it can be seen from their properties, CPS are expected to perform diverse operations not only directed to their cybernetic behaviour, but also to the physical process. Their workflows and the above-mentioned operational modes introduce principal security requirements for CPS, which any security architecture for CPS should be designed to meet:

- *Sensing Security*: as CPS are closely related to the physical processes they are embedded in, the validity and accuracy of the sensing process have to be ensured. Sensing Security needs techniques to authenticate physical stimuli, so that any data measured in the physical processes can be trusted.
- *Storage Security*: once data have been collected and processed, they may be required to be stored over time for future access. Any tampering of these stored data can lead to errors during planning. Storage Security involves developing solutions for securing stored data in CPS platforms from physical or cyber-tampering.
- *Communication Security*: an important aspect of CPS is that they are networked by nature. This does not only allow them to form networks for data fusion and delivery to back-end entities, but also to take coordinated response actions (in both the semi-active and active operational modes). Communication Security needs the development of protocols to secure both inter- and intra-CPS communication from both passive (eavesdroppers) and active (interferers) adversaries.

- *Actuation Control Security* refers to ensuring that no actuation can take place without appropriate authorisation during the semi-active or active modes of operation. The authorisations have to be specified dynamically as the requirements for CPS change over time.
- *Feedback Security* refers to ensuring protection of the control systems in a CPS which provide the necessary feedback for effecting actuation.

The current security solutions focus on data security only, but their effects on estimation and control algorithms have to be studied to provide in-depth defence against CPS tampering [14].

1.3 Engineering Cyber-Physical Systems

Considering the various disciplines involved, CPS engineering is very demanding. Design issues usually arise with the interfaces among the physical and cybernetical components. Also, for a CPS, timeliness, safety and security are equally important as functional correctness. Hence, CPS should be designed holistically, considering all their components and functional properties.

To provide for the above-mentioned properties, they should be designed for verifiability, ensuring correctness of their operation, and ease of validation to check for coherence with their specifications. To enable their design for fostering verification and validation, formal languages and mathematical notations enabling formal proofs (e.g. formal languages and timed automata Agha [15], graphical techniques with the same expressive power as their formal language counterparts Dietz [16], and combinations of conventional CASE methods and statecharts Traoré and Sahraoui [17]) have been used. While enabling formal verification, however, most of these methods lack the versatility of basic constructs and user friendliness. Therefore, graphical formalisms with richer sets of basic constructs have been defined (e.g. CSR/CCSR by Lee et al. [18], GCSR by Ben-Abdallah and Lee [19], TTM/RTTL by Ostroff [20]), while keeping enough "strictness" to enable verifiability. Dedicated state transition automata like CRSM (Shaw [21]) have often been used as basic internal computation models (e.g. POLIS by Balarin et al. [22]).

To avoid exhaustive testing or combinatorial explosion in formal verification, simulation is often used to check the correctness of a system designed or parts thereof. Co-designing systems with temporal limitations also led to the introduction of real-time scheduling strategies into their co-design and co-simulation (e.g. Mooney and Micheli [23]). VHDL is a good example of a specification language suitable for embedded systems, enabling verification and validation, as various verification and validation methods have been devised for VHDL, ranging from formal methods to simulation with fault insertion and combinations thereof.

With the ever increasing complexity of CPS, the traditional development process of manual coding followed by extensive and lengthy testing is becoming inadequate. The main design concern, which first moved from low- to high-level programming languages, recently moved to a higher abstraction level, which relies on automatic or semi-automatic code generators to produce code in traditional programming languages. Examples of these include the Unified Modelling Language (UML) [24], Model-Driven Architecture (MDA) [25] and Model-Integrated Computing (MIC) [26].

When co-designing a CPS, generally three viewpoints must be considered:

1. the external (functional) one, which considers its inputs/outputs and usage scenarios,
2. the internal (behavioural) one, which deals with the definition of usage scenarios, and
3. the definition of system structure—hardware and software architectures together with the mapping of software components onto hardware components and the definition of configurations and reconfiguration scenarios.

1.4   Specification PEARL Approach

PEARL stands for Process and Experiment Automation Real-Time Language, a programming language conceived in the 1990s. PEARL is a standardised programming language [27, 28], developed for programming automation applications for real-time systems. It is one of the rare programming languages allowing the developers to use times explicitly to specify the start times of activities and/or to limit their durations. Like similar third generation high-level languages were developed primarily to suppress complexity. Like many related originally structural languages (e.g. PASCAL, C), it has also been extended for the object-oriented programming paradigm [29]. Its first implementation in the form of a compiler and target platform, consisting of a testbed and real-time operating system was the PEARL90 [30].

Much research has been done on the PEARL90, being extended in two distinct directions:

1. Verifiable PEARL, Safe PEARL, PEARL* in order to enable its formal verification, enhance safety of applications written in PEARL and enable its object orientation.
2. PEARL for Distributed Systems or Multiprocessor PEARL [31] in order to enable systematic design and development of distributed real-time applications.

The basic guidelines and rules being followed in the design of the new PEARL language and methodology:

- co-design language
- language with explicit timing features for determining time constraints
- finite state machine oriented language (easier composition of robust programes)
- safe language (no infinite loops, no pointers, no recursion, …, function cell safety)
- security features (I/O authentication and authorisation)
- possibly an "open language" (addressing the open source community)

Based on its predecessors the language shall build on their best features. Some problematic properties, however, shall be excluded, wiz. explicit task activations, multi-trigger conditions, priorities (that are considered misleading, introducing deadlines instead), etc. On the other hand some additional safety features shall be included, viz. timeout and exception handling, dynamic reconfiguration, secure I/O, etc.

Three ultimate goals shall be followed in the course of defining the new PEARL standard, namely:

- simplicity over complexity,
- inherent real-time ability, and
- conformity to safety integrity and security capability levels.

It shall address the CPS community and shall be an interpreted language, meaning for each platform there will be a virtual machine in the form of a software configuration management executive programme running on the platform and executing PEARL commands.

In this book, the features of the Specification PEARL language and the underlying hardware/software co-design methodology for embedded and CPS are presented and discussed:

- the specification language and graphical notation, which represent hardware/software architectures,
- its timed state transition diagrams, which consistently represent the programme tasks of any real-time application,
- a configuration management mechanism for dynamic system (re-) configuration, and
- co-simulation to verify and validate the anticipated QoS of designs.

Specification PEARL builds on PEARL for Distributed Systems and is meant for programming distributed CPS. It is meant to make them inherently safe "safe by design". In addition it shall include security features in order to address this topic of increasing importance for CPS development.

After the description of the Specification PEARL modelling approach, an interface of the Specification PEARL methodology to UML 2.0 [24] and its extension UML-RT [32] will be presented. Since UML, being a prominent methodology for designing information systems, is also used to design embedded systems, an UML profile for Specification PEARL was defined. Combining both methodologies would enable larger scale Specification PEARL-oriented design of CPS in combination with UML's versatile diagrammatic features. To enable safety and security in the designed systems, a safety and security pattern has been added to the UML-RT profile for Specification PEARL.

There is a number of Quality of Service (QoS) criteria pertaining to CPS. Throughout this book, safety and security of CPS and their ability to be licensable for these properties will be emphasised in order to guarantee their concordance with appropriate standards. The Specification PEARL methodology will be evaluated against the standard IEC 61508, [33] which includes the activities necessary to be carried out for a safety-related system from the start of its design project until the end of its life cycle. Like safety, security is also an issue rapidly gaining importance for CPS. To be effective, it must be dealt with already during the design phase too. Throughout this book the IEC 62443 [10, 11] standards will be considered as a reference for CPS security.

References

1. Bass, M., Christensen, C.: The future of the microprocessor business. Spectr. IEEE **39**(4), 34–39 (2002). doi:10.1109/6.993786
2. Lee, E.A.: Cyber physical systems: design challenges. In: Proceedings of the 2008 11th IEEE Symposium on Object Oriented Real-Time Distributed Computing, ISORC '08, pp. 363–369. IEEE Computer Society, Washington, DC, USA (2008). doi:10.1109/ISORC.2008.25. http://dx.doi.org/10.1109/ISORC.2008.25
3. Adelstein, F., Gupta, S.S.L.: Fundamentals of Mobile and Pervasive Computing. McGraw-Hill Professional Engineering, McGraw-Hill (2005). http://books.google.si/books?id=IhMfAQAAIAAJ
4. Lee, E.A.: Computing needs time. Commun. ACM **52**(5), 70–79 (2009). doi:10.1145/1506409.1506426. http://doi.acm.org/10.1145/1506409.1506426
5. Gumzej, R.: Real-time Systems' Quality of Service. Springer, Dordrecht (2010)
6. Institution, B.S., for Standardization, I.O.: Implementation of ISO/IEC 13236: Information Technology: Quality of Service: Framework. British Standards Institution. http://books.google.si/books?id=mpkgHAAACAAJ (1996)
7. ANSI/ISA 99.01.01-2007: Security for industrial automation and control systems part 1: terminology, concepts, and models. http://webstore.ansi.org/RecordDetail.aspx?sku=ANSI (2007)
8. ANSI/ISA 99.01.02-2009: Security for industrial automation and control systems: establishing an industrial automation and control systems security program. http://webstore.ansi.org/RecordDetail.aspx?sku=ANSI (2009)
9. ANSI/ISA 99.03.03-2013: Security for industrial automation and control systems part 3–3: system security requirements and security levels. http://webstore.ansi.org/RecordDetail.aspx?sku=ANSI (2013)
10. IEC TS 62443-1-1:2009, Industrial communication networks - network and system security - part 1–1: terminology, concepts and models. https://webstore.iec.ch/publication/7029 (2009)
11. IEC 62443-2-1:2010, Industrial communication networks - network and system security - part 2–1: establishing an industrial automation and control system security program. https://webstore.iec.ch/publication/7030 (2010)
12. Stouffer, K., Falco, J., Scarfone, K.: Guide to industrial control systems (ics) security. Technical Report, NIST (2011). http://csrc.nist.gov/publications/nistpubs/800-82/SP800-82-final.pdf
13. Kephart, J.O., Chess, D.M.: The vision of autonomic computing. IEEE Comput. **36**(1), 41–50 (2003)
14. Cárdenas, A.A., Amin, S., Sastry, S.: Research challenges for the security of control systems. In: Proceedings of the 3rd Conference on Hot Topics in Security, HOTSEC'08, pp. 6:1–6:6. USENIX Association, Berkeley, USA (2008). http://dl.acm.org/citation.cfm?id=1496671.1496677
15. Agha, G.: The structure and semantics of actor languages. In: de Bakker, J.W., de Roever, W.P., Rozenberg, G. (eds.) Foundations of Object-Oriented Languages, pp. 1–59. Springer, Berlin (1991)
16. Dietz, C.: Action diagrams. In: M. Maranzana (ed.) Proceedings of the IFAC/IFIP Workshop, 15–17 September 1997, Real-Time Programming 1997. Lyon, France, Elsevier Science 1998 (1997). http://csd.informatik.uni-oldenburg.de/pub/Papers/cd97-a.ps.gz An abstract is available on-line

17. Traoré, I., Sahraoui, A.: A multiformalism specification framework with statecharts and vdm. In: 22nd IFAC/IFIP Workshop on Real-Time Programming (WRTPÕ97), pp. 63–68 (1997)

18. Lee, I., Davidson, S., Gerber, R.: Communicating Shared Resources: A Paradigm for Integrating Real-time Specification and Implementation. GRASP LAB: General Robotics and Active Sensory Perception Laboratory. University of Pennsylvania, School of Engineering and Applied Science, Department of Computer and Information Science. http://books.google.si/books?id=_QttuAAACAAJ (1991)

19. Ben-Abdallah, H., Lee, I.: A graphical language for specifying and analyzing real-time systems. Integr. Comput.-Aided Eng. 5(4), 279–302 (1998). http://dl.acm.org/citation.cfm?id=1275802.1275805

20. Ostroff, J.: A visual toolset for the design of real-time discrete-event systems. IEEE Trans. Control Syst. Technol. 5(3), 320–337 (1997)

21. Shaw, A.: Communicating real-time state machines. IEEE Trans. Softw. Eng. 18(9), 805–816 (1992). http://doi.ieeecomputersociety.org/10.1109/32.159840

22. Balarin, F., Chiodo, M., Giusto, P., Hsieh, H., Jurecska, A., Lavagno, L., Passerone, C., Sangiovanni-Vincentelli, A., Sentovich, E., Suzuki, K., Tabbara, B. (eds.): Hardware-Software Co-design of Embedded Systems: The POLIS Approach. Kluwer Academic Publishers, Norwell (1997)

23. Mooney Iii, V.J., De Micheli, G.: Hardware/software co-design of run-time schedulers for real-time systems. Des. Autom. Embedded Syst. 6(1), 89–144 (2000)

24. OMG: Unified modeling language (uml) resource page. http://www.uml.org/ (2015)

25. OMG: Mda - the architecture of choice for a changing world. http://www.omg.org/mda/ (2015)

26. Sztipanovits, J., Karsai, G.: Model-integrated computing. Computer 30(4), 110–111 (1997). doi:10.1109/2.585163

27. 66253, Part 1: Basic pearl. Technical Report, DIN (1981)

28. 66253 Part 2: Full pearl. Technical Report, DIN (1982)

29. Frigeri, A.H., Pereira, C.E., Halang, W.A.: An object-oriented extension to pearl90. In: ISORC, pp. 265–274 (1998)

30. Pearl - process and experiment automation realtime language. http://www.pearl90.de/ (2014)

31. 66253 Part 3: Pearl for distributed systems. Technical Report, DIN (1989)

32. Herzberg, D.: Uml-rt as a candidate for modeling embedded real-time systems in the telecommunication domain. In: France, R., Rumpe, B. (eds.) ÇUMLßÕ99- The Unified Modeling Language, Lecture Notes in Computer Science, vol. 1723, pp. 330–338. Springer, Berlin (1999). doi:10.1007/3-540-46852-8_24. http://dx.doi.org/10.1007/3-540-46852-8_24

33. 65A, I.S.: Functional safety of electrical/electronic/programmable electronic safety-related systems. Tech. Rep. IEC 61508, The International Electrotechnical Commission, 3, rue de Varembé, Case postale 131, CH-1211 Genève 20, Switzerland (1998)

Chapter 2
Specification PEARL Language

2.1 Extending PEARL for Distributed Systems

Since the complexity of current automation and real-time processing tasks requires the programming of distributed, fault-tolerant multiprocessor systems, the developers of PEARL have decided to extend PEARL with constructs for the programming of multiprocessors. Thus, Multiprocessor PEARL or PEARL for Distributed Systems, viz. DIN 66253, Part 3 [1], was defined as an over-layer on PEARL, and enhanced the language with constructs for the abstract descriptions of hardware and software architectures. These enabled real-time embedded systems to be co-designed in order to increase their quality of service, in particular their predictability and dependability. While not being translated into machine code, these constructs are mainly used as directives for system programs (e.g. real-time operating systems, configuration management programs, etc.) instead. Hence, Multiprocessor PEARL has further been extended in the form of a co-design methodology into the Specification PEARL language and methodology with the following properties:

- constructs to describe hardware configurations,
- constructs to describe software configurations,
- constructs to specify communication and its characteristics (peripheral and process connections, physical and logical connections, transmission protocols) as well as
- constructs to specify both conditions and methods of carrying out dynamic reconfigurations in cases of failure.

Furthermore, Specification PEARL has the following characteristics, usually required for specification languages:

- abstraction, i.e. insignificant details are suppressed, the conceptual world of the application domain is supported, and no implementation is referred to,
- application concepts and structures, relations and sequences are easily recognisable,
- easy readability, but nevertheless precise notation,

© Springer International Publishing Switzerland 2016
R. Gumzej, *Engineering Safe and Secure Cyber-Physical Systems*,
Studies in Computational Intelligence 632, DOI 10.1007/978-3-319-28905-2_2

- provision for unambiguous and complete descriptions of requirements and design,
- support for effective communication between clients, designers and users about the systems to be developed,
- possibility of easily extending specifications into executable prototypes,
- inclusion of appropriate real-time executive and dynamic re-configuration management programs and
- systematic integration of the specification method into the entire development process.

Specification PEARL extends PEARL for distributed systems to enable the specification of asymmetrical architectures as well as towards a more distinctive description of systems' communication interfaces and intelligent peripheral devices. Along with the textual man-readable specification language, graphical constructs with the same properties have been defined as basis for an appropriate CASE environment. Within the environment behavioural modelling (of program tasks) by timed state transition diagrams is supplemented. The output models (virtual machines), representing the target systems' hardware and software architectures as well as application program prototypes are subject to verification and validation. As they can be checked for correctness, consistency and coherency, the methodology provides a verification phase preceding the validation phase, where a system's coherence with the prescribed functional, temporal as well as safety and security requirements is checked.

Herewith, a methodology is defined, enabling systematic design of the structure as well as the behaviour of the designed system. Its benefits are the standard-based user-readable syntax, which can serve as input of compilers, configuration managers or loaders, the ability to model a system's dynamic behaviour, which is suitable for validation by simulation, and summarising all this, the ability to check a system's feasibility before implementing it. The methodology is presented in detail in the next chapter.

In the sequel, the Specification PEARL language is presented, followed by the description of its associated CASE environment with its program libraries, to be used in the design, verification, validation and deployment phases. The modelling technique based on timed state transition diagrams is presented to demonstrate, how program tasks are formed. Finally, an example of a typical usage scenario and an existing prototype of a distributed hard real-time system [2] is considered as a case study.

2.2 Specification PEARL Notation

A system architecture specification consists of DIVISIONs, which describe different associated layers of the system design in considerable detail (e.g. Fig. 2.1):

STATION   processing node(s) hardware description,
CONFIGURATION   software unit(s) description,
NET   network interconnection(s) description,

```
ARCHITECTURE;                          CONFIGURATION;
 STATIONS;                               COLLECTION KP_WS;
  NAMES: KP;                              PORTS KP_TP1-lin, KP_TP2-lin;
   PROCTYPE: MC68370 AT 20 MHz;           CONNECT KP_WS.KP-TP1_lin INOUT TP1-WS.TP1_KP_lin
   WORKSTORE: SIZE 65536 SPACE 0 - 'FFFF'B4    VIA KP.KP_IO;
    READ/WRITE WAITCYCLES 1;              CONNECT KP_WS.KP_TP2_lin INOUT TP2_WS.TP2_KP_lin
   WORKSTORE: SIZE 32768 SPACE 0 - '7FFF'B4    VIA KP.KP_IO;
    READONLY WAITCYCLES 1;                COLEND;
   INTERFACE: KP_IO (DRIVER: KPINOUT;
    DIRECTION: INPUT; SPEED:20971520 BPS;  COLLECTION TP_WS;
    UNIT:FIXED);                           PORTS S1, TP1_KP_lin;
   STATEID: (NORMAL, CRITICAL);           CONNECT TP1_WS.S1 IN VIA TP1.S1;
   STATIONTYPE: KERNEL;                    CONNCET-TP1_WS.TP1_KP_lin INOUT
   SCHEDULING: EDF;                         KP_WS.KP_TP1_lin VIA TP1.TP1_IO;
   MAXTASKS: 20;
   MAXSEMA: 5;                            MODULES TP1_WS_M1;
   MAXEVENT: 15;                           EXPORTS(Side 1);
   MAXEVENTQ: 5;                            TASK Side1
   MAXSCHED:30;                             TRIGGER PORT S1;
   TICK: 1E-3 SEC;...                       DEADLINE 100;
                                           TASKEND;
 SYSTEM;                                  MODEND;
  NAMES: KP;                             COLEND;...
   KP.KP_IO INOUT;
   NAMES: Sensor 1;                      CONFEND;
    Sensor1.S1 OUT;                      ARCHEND;
   NAMES: Sensor 2; ...
   NAMES: TP1;                           NET;
    TP1.S1 IN;                            KP.KP_IO <-> TP1.TP1_IO;
   TP1.TP1_IO INOUT;                      KP.KP_IO <-> TP2.TP2_IO;
   NAMES:TP2; ...                         TP1.TP1_IO<-> Sensor1.S1;
 SYSEND;                                  TP2.TP2_IO<-> Sensor2.S2;
                                         NETEND;
```

Fig. 2.1 An example of a textual architecture description expressed in specification PEARL

SYSTEM SW/HW   interface(s) description and
PERIPHERAL   intelligent peripheral device(s) description.

Since contemporary specification formalisms use graphical notations, graphical constructs with the same semantics as the textual BNF-based descriptions were defined as basis for the associated CASE environment. An overview of Specification PEARL's textual syntax and graphical notation is given in Appendix A and Appendix B, respectively.

2.2.1 Hardware Configuration

In the STATION division a system's processing nodes (*stations*) are introduced stating their most important characteristics. Stations are treated as black boxes with connections through their interfaces. To allow for multiprocessor nodes, a *composite station* is defined to be a set of stations, which are logically and physically strongly connected (i.e. they share the same housing or at least the same connections with other stations and/or intelligent peripheral devices). The basic components of a station are its processors (*proctypes*), working storages (*workstores*) and different types of *devices*. There may be multiple stations in a system, so each one of them is uniquely identified. Each station in a system maintains its *state* information.

There are several types of stations. The default type is the BASIC station, which stands for a general-purpose processing node. To be able to describe asymmetrical architectures, two additional types of processing nodes have been defined: TASK for pure application task execution and KERNEL for real-time operating system execution. Since for CPS intelligent peripheral devices are very important, the PERIPHERAL station type was introduced to represent this kind of stations.

Besides the before-mentioned general attributes, stations may also have additional attributes depending on the station's type. A multiprocessor node is characterised by the "PART OF" attributes of its constituent processing nodes. Kernel stations have properties, which are specific to them and are relevant to software designers (e.g. scheduling strategy, maximum number of active tasks, maximum number of synchronisers, events, queued events and schedules supported, real-time clock resolution etc.).

Processor's (PROCTYPE) properties are its unique ID and speed descriptor, which indicates the clock generators' frequency. Although this information may seem irrelevant at this point, one may choose to drive processors with different frequencies, which affects their processing speed. Hence, for a profiler or schedulability analyser this information is crucial to estimate the actual execution times of the individual instructions and tasks.

Work-stores (WORKSTORE) are described by their capacities and memory maps (showing the purpose of different memory areas). The wait-cycles, associated with the individual work-store areas, may also be specified (on-chip, random access or read-only memories usually have different access times). This information is used by compilers to determine the maximum execution times of tasks, being loaded to these memory areas, or the time required to access their data during execution.

Devices (DEVICE) are identified by IDs (like stations, but they may be assigned a logical name for easier reference). The device types may vary and have different attributes assigned depending on their nature. Currently, INTERFACE, TIMER and SHARED variable device types are foreseen. The use of standard devices is supported by the generic device specification.

Any net topology of a distributed system can be described by point-to-point connections. Hence, a NET division describes the physical connections between the stations of a system by their logical names and directions.

A SYSTEM division encapsulates a hardware description and the assignment of all relevant symbolic names to hardware devices, where the described components from the station and net divisions are referenced by their IDs.

A PERIPHERAL division provides the details about the intelligent peripheral devices attached to a system. The peripheral devices are identified by their IDs. Their connections to the stations of the system are described by the logical names of the interfaces, they are attached to, as well as the attributes of the interfaces used for communication (e.g. direction of data flow, protocol used and any additional signals which may be necessary for communication). To support schedulability analysis, every signal from a peripheral device can be associated with its minimum inter-arrival time.

2.2.2 Software Configuration

The CONFIGURATION division deals with a system's software architecture. The biggest program part associated with a STATION is a COLLECTION. Collections are composed of modules (MODULE) of tasks (TASK) which communicate through the respective collection's ports (PORT). Each program part has its unique name for reference.

Modules are mainly meant for information hiding and sharing among bigger collections' parts. Hence, they are further described by their IMPORTS and EXPORTS, where it is stated which data structures and tasks are shared with other modules.

Tasks are described by their trigger conditions and response times. Task alternatives may be provided in order to increase fault tolerance by graceful degradation through task scheduling—an alternative with shorter run-time or longer response time may be scheduled to maintain a schedule's feasibility.

Collections are software architecture units, being associated with stations. They are loaded to stations by re-configuration management programs, being in charge of the initial loading and starting the initial tasks. It is possible to specify under which conditions certain collections are to be removed from a station and which collections are to be loaded instead. The initial configuration as well as reconfiguration conditions are station state dependent. The latter is maintained by the already mentioned re-configuration management programs or *configuration managers*.

Collection's ports are used for inter-collection (inter-station) communication. Hence, they are associated with appropriate station interfaces. The connections between the ports of collections are described by their directions and line attributes. Port's line attributes state which connections are always used (VIA attribute) and which ones can be chosen from a preference list based on the PREFER attribute (e.g. when using multiple different interfaces for the same communication line to increase line robustness).

2.3   Specification PEARL CASE Environment and its Program Libraries

Most of the design methodologies do not consider the target platform—hardware architecture and operating system—and only some of them have their associated CASE environments. Those, which do and produce executable code, use off-the-shelf operating systems with the corresponding tools being strictly bound to the target environments, from which they also inherit their strengths and weaknesses, in particular limitations in their capabilities, connectivity and real-time ability.

From this point of view, it was meaningful to apply a holistic approach also to building the Specification PEARL CASE environment. Since the PEARL language already includes appropriate calls to the operating system, an appropriate real-time operating system was developed for it. To provide for the appropriate parameterisation of the operating system and appropriate consideration of the system architecture and interfaces a hardware abstraction layer was built in form of a basic executive and data interchange program. Two libraries were built for Specification PEARL and bundled with its CASE environment [3].

The first program library represents the real-time operating system, which can be easily combined with the designed application code. A rich set of system calls supporting real-time operation was foreseen, based on past experience with the RTOS-UH [4], which was primarily bound with the PEARL90 [5] compiler to build and execute real-time applications on a proprietary platform. The HaRTOS [2] real-time operating system supports the PEARL's tasking model and system calls as well as the deadline-driven scheduling strategy and is thus capable of ensuring hard real-time operation. It is programmed in C and can, thus, be compiled for any hardware platform. HaRTOS resources are pre-determined (e.g. maximum number of tasks, synchronisers, signals, events or queued events) by adequately configuring the station's parameters.

The second program library—the Configuration Manager (CM)—is meant tó be used as the main executive program at each station. It represents a hardware abstraction layer that is, as configured by a hardware architecture model, mainly used to define the structure and interfaces of each station. It initiates the execution of the initial task and monitors the station state. In case the station state changes, it performs the pre-specified re-configuration actions, which mainly comprise fail-safe ending of the current execution and fast scenario switching to another (collection of) task(s). HaRTOS, is accessible through a proprietary CM communication channel. The rest of the station's communication channels are used as pre-configured by the hardware architecture description through appropriate CM data interchange methods.

In the Specification PEARL CASE environment, hardware and software architectures are designed conjunctly—one may start the design from either point of view and associate them later on. A completeness and consistency check is done in order to ensure model completeness and parameter consistency. The Specification PEARL CASE environment, which encompasses modelling and co-simulation tools, also enables cross-development for the specified hardware architectures by cross-compilation of its target platform models. The environment supports the Specification PEARL project life-cycle (c.p. Fig. 3.1), as presented in the next chapter.

2.3.1 Configuration Manager

At each processing node (station) execution starts by initiating its configuration manager (CM) which, in turn, loads the initial collection by triggering the latter's initialisation tasks. In stations without a real-time operating system, the collection's main task is started and delegated control to by the CM, whereas otherwise the CM acts as a front-end to the operating system, and uses appropriate system calls and system ports to transfer system requests to/from RTOS-enabled nodes to schedule the collection's tasks.

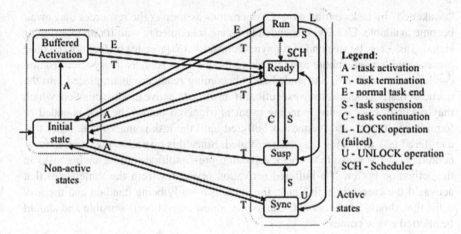

Fig. 2.2 Task model of RTOS

Besides local execution, the CM is also responsible for communication with other stations, and for co-operation among the tasks of the same collection. Hence, it must establish port-to-port connections through the interfaces of the station and provide for task synchronisation through HaRTOS. Synchronisation and system service requests are serviced on the same station, in case the station is configured to run the operating system. Otherwise, these requests are delegated to the corresponding (KERNEL) station through a proprietary port. The functions of the CM are described in detail in Appendix Є.

2.3.2   Operating System

The implementation of the HaRTOS Real-Time Operating System (HaRTOS) is primarily oriented at PEARL. It supports PEARL's task model (cp. Fig. 2.2) and the system calls defined by standard PEARL with a few enhancements. The detailed description of the RTOS library's API and of its system services is given in Appendix D. In part, the implementation of HaRTOS also addresses the already mentioned CM.

According to Fig. 2.2, tasks may be active or inactive (dormant). They are activated by a (scheduled) activation (A), and return to inactive state after a normal task end (E) or task termination (T). When active, tasks may be running or awaiting their *run* conditions fulfilment, in the queue of *ready* tasks or in one of the suspended states. Ready tasks are ready to run, however, they are awaiting their turn—their queue is maintained by the HaRTOS scheduler (SCH). Tasks may be suspended for two reasons—awaiting resource allocation or synchronisation. The tasks waiting for resources are put into the *Susp* state by task suspension (S) operations and are "awakened" by task continuation (C) operations as soon as the resources they await become available. Usually, suspended tasks are scheduled to wait for an interrupt or signal. Tasks can be suspended for synchronisation (*Sync* state) by locking (L) their presence with a semaphore and unlocking them (U) as soon as the synchronisation conditions are fulfilled. Suspended tasks becoming ready are again placed into the queue of ready tasks awaiting execution. If an already active task is activated, which may occur in case a corresponding event is triggered before the task is ended or terminated, its activation request is buffered until the task is inactivated. After that a buffered task activation may be performed. Since this operation may cause "race-conditions", it is sensible to check activation pre-conditions before saving tasks in the activation buffer. If a buffered activation originates from the same event that activated the currently active task, the event is already being handled and the new activation should not be stored. Otherwise, a new activation is sensible and should be allotted a new context.

2.4 Specification PEARL Behavioural Model

The program tasks of an application represent the processes in a running system. They are mainly characterised by their activation conditions and timing limitations as well as by their adherence to collections and modules. This information is sufficient to build a coarse program model, but it is not enough to determine its feasibility. Therefore, timed state transition diagrams (TSTD) were introduced to represent them [6]. Their synchronisation and inter-communication are realised by calls to the configuration manager and the real-time operating system of the station, respectively.

   TSTD are hierarchical finite state automata consisting of

- start states: task activation conditions and initialisation actions,
- transient states: atomic activities with possibly predictable duration,
- super-states: non-atomic activities—hierarchical decomposition of working states, and
- final states: finalisation actions.

The connections between states represent the progress of tasks from start to final states. In each state they spend some time, so their progression through them also represents their progression in time. All connections among states are local (i.e. bound to one task). In every state some actions, whose execution is considered atomic, may be executed. These actions also trigger the continuation pre-conditions of the states. Intertask co-operation is enabled—by appropriate system calls to the operating system through the configuration manager. Operating system and configuration manager are accessible through their APIs and behave in concordance with the pre-specified system configuration.

   The formal representation of a program configuration is the union of the tasks' models:

$$M^* = \cup_{i=0}^{n} M_i$$

Any task is represented by an eight-tuple:

$$M_i = (S, \Sigma, V, I, \delta, \varepsilon, \tau, E)$$

with the following components:

1. S_i: set of start states (there may be more, since there may be multiple trigger conditions for a single task),
2. Σ_i: set of input symbols (states trigger conditions),
3. V_i: set of all states,
4. I_i: set of time intervals (every state may be associated with one; $i_i = (v_i, t_i)$ where for each $i_i \in I_i$, $v_i \in V_i$ is the current state, $t_i \in I_i$ is the predefined time-out interval),
5. δ_i: state transition function,
6. ε_i: semantic state function (represents actions which change the internal state of the station and enable inter-task synchronisation and communication),
7. τ_i: transition function in case of time-out,
8. E_i: set of final states.

Any state is characterised by the following data:

- state type (start, transient, supet or final state),
- pre-condition for the state's execution (trigger condition for a start state),
- time-out condition (shortest, mean and maximum execution time),
- time-out action (state to which execution is diverted in case the time-out occurs),
- connection to the next state(s) in case the continuation condition(s) are fulfilled on-time and task execution within the state is successful,
- activities carried out within the framework of this state (designer-defined actions and system calls).

Start states represent different task entry points. Transient states represent states within the execution of a task where a task resides and does some action(s) awaiting the pre-condition(s) of its successor state(s) to be fulfilled. Only initial tasks' start states have the possibility of explicit (on-demand) activation. To enter other task states the following types of pre-conditions must be fulfilled:

- external events: int(number), representing interrupts (discrete signals),
- internal events: sig(identifier), representing signals,
- timers: timer(at, every, during), representing timer signals,
- general conditions: cd(expression), i.e. expressions returning Boolean results from the evaluation of internal station/program states or data structures of the operating system.

Task execution progresses from state to state upon fulfilment of a pre-condition of a successor state. Upon successful completion of the execution of the task finalisation actions within its final state control is returned to its initial/start state(s). Upon fulfiling the pre-condition of a super-state, control is automatically transferred to the start state of its sub-model. When the final state of the sub-model is reached, control is returned to the super-state awaiting continuation pre-conditions.

Any transient state may be allotted a minimum ($minT$) and maximum ($maxT$) time-frame for its execution. The time-out condition is set to the maximum time-frame at the beginning of each state's execution. If the time-out condition is not reached before the condition to proceed to the next state is fulfilled, the corresponding connection for successful continuation is followed. If a minimum time-frame is foreseen, the continuation conditions are not checked before the specified time has elapsed. In case a time-out occurs, an appropriate on-timeout action is executed. If a time-out occurs and no on-timeout action is specified, an error is raised (and logged in the co-simulation).

The activities within a state are a set of actions, which are carried out while the task is in this state. It is assumed that the actions form a single block of program statements including system calls to the operating system and/or configuration manager, around which the control structure is formed while transforming the state chart to program code. The designer's estimates of their minimum and maximum execution times are the basis for setting the respective time-frames for the state.

The system calls, performed within a state, may change station state and hereby affect the execution of the current or another task. They may claim and release resources. They may make inquiries on the internal state of the station or change the internal state of the station. They may synchronise task execution. They may transfer task data to another task/collection by utilising appropriate system calls to the CM with references to appropriate ports and interfaces as pre-configured by the hardware/software architecture.

Hence, we may conclude that, although task state transition diagrams determine the course of individual task execution, the overall task execution is controlled by the operating system and the CM. In case a context switch is necessary, the current task's state is saved with the task's context and re-established, when task execution continues.

2.4.1 Task-Forming Rules

The role of a "task" is the same in the Specification PEARL methodology as it is in the associated programming language PEARL [7, 8]—any procedure, which needs to be carried out within a given time-frame, is a task. Therefore, we can generally say that a task is the greatest program unit, to which a maximum execution time or a deadline can be assigned.

The problem in trying to break task operations down into states is that simple tasks have just three states, viz. start, working (transient) and final. New states are only introduced (1) if a time-limited atomic (sub) operation is identified, (2) if synchronisation or communication between tasks is necessary or (3) to define branching into different continuation paths depending on the pre-conditions of successor states. The following criteria were selected to form task states:

- a state represents a single logical activity, which is only dependent on its pre-conditions and whose execution time can be determined or predicted,
- any task must have at least one start state and one or more transient-/super- and final states,
- to facilitate good decomposition, a complex state shall be broken down into simpler states by introducing a super-state and defining its state-transitions in a sub-diagram.

2.4.2 Translation from Timed State Transition Diagrams to Program Tasks

When deploying the system model, program code is automatically generated from TSTD diagrams. The general form of task prototypes, obtained from task TSTDs, is shown in Fig. 2.3.

TSTD task models are translated to program tasks in two forms:

1. target-platform-oriented, as they can be compiled by a corresponding compiler and executed on the specified hardware architecture, and
2. simulation-oriented, as they can be used and interpreted by co-simulation in a simulation environment.

The main difference between the two forms is the way external events are handled. In the first case, they are generated by the environment and handled as hardware interrupts (by appropriate device drivers), whereas in the second one, they are generated in the co-simulation environment and handled as software signals (by stub device drivers). In both cases, they are handled by the station's CM and operating system.

2.5 Case Study—Railroad Crossing

This is a well-known example used throughout the literature on formal methods for the domain of real-time systems. It is a good example to demonstrate the principles of the methodology described in this book. Here, we consider a crossing with two tracks, although there could be more. There are two sensors, S1 and S2, guarding the railroad crossing one on each side (Fig. 2.4). The railway barriers are closed and opened upon receipt of the corresponding signals.

For this application, the following demands and restrictions hold:

- Safety: if a train is in the crossing, the railway barrier is closed,
- Responsiveness: the railway barriers are open most of the time,

```
MODULE module_name;                                      #include "module_name.h"
SYSTEM;
! interrupts, signals and system variables definitions   /* interrupts, signals and system variables definitions */
PROBLEM;

task_name : PROCEDURE (state_id REF INT);                void task_name(int &state_id;) {
  DCL timeout BIT;                                          bool timeout;
                      /* initialisation of all global structures  */
timeout:='0'B1; ! timeout indicator                      timeout=false; /* timeout indicator */
WHILE '1'B1 REPEAT                                        while (1) {
  CASE state_id                                            switch (state_id) {
    ALT (0) ! START state:                                  case 0: { /* START state: */
      IF timeout EQ '1'B1 THEN                                if (timeout) {
                      /* perform OnTimeout=action(s);   */
      state_id:=0; timeout:='0'B1; NEXT state_id;              state_id=0; timeout=false; Next(state_id);}
      ELSE                                                     else {
                      /* RESUME task after fulfillment of the  trigger conditions; */
      timeout:='1'B1;                                          timeout=true;
      DELAY maxT;                                              Delay(maxT);
                      /* perform the appropriate start states  Action=statements; */
      /* check if any of the next working / super / end states'  pre-conditions are fulfilled; */
                /* if they are fulfilled, set timeout to false and  set the state_id variable accordingly; */
      NEXT state_id;                                            Next(state_id);
      FIN;                                                   } }
    ALT (1) ! for a WORKING state:                         case 1: { /* for a WORKING state: */
      IF timeout EQ '1'B1 THEN                                if (timeout) {
                      /* perform OnTimeout=action(s);   */
      state_id:=0; timeout:='0'B1; NEXT state_id;              state_id=0; timeout=false; Next(state_id);}
      ELSE                                                     else {
      DELAY minT;                                              Delay(minT);
      timeout:='1'B1;                                          timeout=true;
      DELAY maxT;                                              Delay(maxT);
                      /* perform Action=statements;   */
      /* check if any of the next working / super / end states'  pre-conditions are fulfilled; */
                /* if they are fulfilled, set timeout to false and  set the state_id variable accordingly; */
      NEXT state_id;                                            Next(state_id);
      FIN;                                                   } }
    ALT (2) ! for a SUPER state:                           case 2: { /* for a SUPER state: */
      IF timeout EQ '1'B1 THEN                                if (timeout) {
                      /* perform OnTimeout=action(s);   */
      state_id:=0; timeout:='0'B1; NEXT state_id;              state_id=0; timeout=false; Next(state_id);}
      ELSE                                                     else {
                      /* set the state_id variable to the start state of  the super state's sub-diagram; */
      NEXT state_id;                                            Next(state_id);
      FIN;                                                   } }
    ALT (3) ! for a SUPER state (as addition) - "return to" state:  case 3: { /* for a SUPER state (as addition) - "return to" state: */
                      /* set the state_id variable to the next state  */
      /* for a SUPER state (the sub-diagram states numbered  consecutively) */
      NEXT state_id;                                          Next(state_id); }
    ALT (n) ! END state:                                   case n: { /* END state: */
      DELAY minT;                                             Delay(minT);
                      /* perform Action=statements;   */
      state_id:=0;                                            state_id=0;
                      /* reset task state (state_id = super_state + 1)  in case of a return from a sub-diagram */
      NEXT state_id;                                          Next(state_id);
      timeout:='0'B1; ! reset timeout indicator                timeout=false; /* reset timeout indicator */ }
    FIN;                                                   }
  END;                                                   }
END;                                                   }

MODEND;
```

* DELAY and NEXT are there for simulator control. DELAY represents a "busy wait" while NEXT instructs the simulator to dispatch ("preemption point").

Fig. 2.3 Representations of a TSTD in PEARL and annotated C

under the following pre-conditions:

- a train shall not arrive (in the protected area) before the previous one has left,
- if a train arrives from the left, it leaves on the right and vice-versa,
- *dmin* is the minimum time for a train to arrive in the railroad crossing after the sensor detects it,
- *dopen*, *dclose* are maximum opening/closing times of the railway barrier.

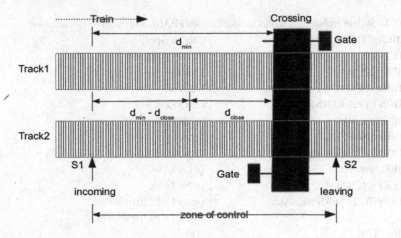

Fig. 2.4 Scheme of the railroad crossing problem

An algorithmic solution to this problem reads as follows:

> **var** x ranges over Tracks;
> initial values:
>
> Deadline(x) := ∞ **forall** x ∈ Tracks Dir := open;
>
> **forall** x in parallel **repeat**
> **block**
>
> **if** TrackStatus(x) = coming **and** Deadline = ∞ **then**
>   Deadline(x) := $dclose + dmin$;
> **endif**
> **if** TrackStatus(x) = empty **and** Deadline < ∞ **then**
>   Deadline := ∞;
> **endif**
> **if** Dir = open **and not**(SafeToOpen) **then**
>   Dir := close;
> **endif**
> **if** Dir = close **and** SafeToOpen **then**
>   Dir := open;
> **endif**
>
> **endblock**

The solution of the problem is represented by an asymmetrical system with three stations (Table 2.1). It consists of a control node (KP running the operating system) and two task processing nodes (TP1 and TP2) servicing the signals coming from two sensors—one at each side of the gate. The task processors are linked to the control node and accept input signals from their respective sensors Sensor 1 (S1) and Sensor 2 (S2). The sensors S1 and S2 have to be activated one after the other for an adequate passing of a train.

Table 2.1 System architecture model in Specification PEARL and internal representation

ARCHITECTURE:	[Station types]
STATIONS:	COMPOSITE = 0
NAMES: KP;	BASIC = 1
STATEID: (NORMAL);	TASK = 2
STATIONTYPE: KERNEL;	KERNEL = 3
NAMES: Sensor1;	PERIPHERAL = 4
STATEID: (NORMAL);	[Stations]
STATIONTYPE: PERIPHERAL;	KP = KERNEL
NAMES: Sensor2;	TP1 = TASK
STATEID: (NORMAL);	TP2 = TASK
STATIONTYPE: PERIPHERAL;	Sensor1 = PERIPHERAL
	Sensor2 = PERIPHERAL
NAMES: TP1;	[KP]
STATEID: (NORMAL);	Step = 10
STATIONTYPE: TASK;	[TP1]
NAMES: TP2;	Step = 10
STATEID: (NORMAL);	Supervisor = KP
STATIONTYPE: TASK;	[Collections.TP1]
STAEND;	Name1 = TP1_WS
NET:	
KP.KP_TP1_lin INOUT TP1.TP1_KP_lin;	[TP1_WS]
KP.KP_TP2_lin INOUT TP2.TP2_KP_lin;	State = NORMAL
TP1.S1 IN;	[Tasks.TP1_WS]
TP2.S2 IN;	Name1 = Sensor1
TP1.TP1_KP_lin INOUT KP.KP_TP1_lin;	[Sensor1]
TP2.TP2_KP_lin INOUT KP.KP_TP2_lin;	Trigger = PORT S1
NETEND;	Deadline = d_Sensor
SYSTEM:	Init = Sensor1.ini
NAMES: KP;	[TP2]
KP.KP_TP1_lin INOUT TP1.TP1_KP_lin;	Step = 10
KP.KP_TP2_lin INOUT TP2.TP2_KP_lin;	Supervisor = KP
NAMES: Sensor1;	[Collections.TP2]
NAMES: Sensor2;	Name1 = TP2_WS
NAMES: TP1;	[TP2_WS]
TP1.S1 IN;	State = NORMAL
TP1.TP1_KP_lin INOUT KP.KP_TP1_lin;	[Tasks.TP2_WS]
NAMES: TP2;	Name1 = Sensor2
TP2.S2 IN;	[Sensor2]
TP2.TP2_KP_lin INOUT KP.KP_TP2_lin;	Trigger = PORT S2
SYSEND;	Deadline = d_Sensor
CONFIGURATION:	Init = Sensor2.ini

(continued)

Table 2.1 (continued)

COLLECTION KP_WS	[Sensor1]
PORTS KP_TP1_lin,KP_TP2_lin;	Step = 1
CONNECT KP_WS.KP_TP1_lin INOUT TP1_WS.TP1_KP_lin VIA	[Collections.Sensor1]
KP.KP_TP1_lin;	Name1 = Sensor1_WS
CONNECT KP_WS.KP_TP2_lin INOUT TP2_WS.TP2_KP_lin VIA	[Sensor2]
KP.KP_TP2_lin;	Step = 1
COLLECTION TP1_WS	[Collections.Sensor2]
MODULES	Name1 = Sensor2_WS
TP1_WS_M1	[Port Map]
EXPORTS (Sensor1)	Port1 = KP.KP_TP1_lin< − >TP1.TP1_KP_lin
TASKS	Port2 = KP.KP_TP2_lin< − >TP2.TP2_KP_lin
Sensor1 (TRIGGER PORT S1,DEADLINE d_Sensor)	Port3 = Sensor1.S1< − >TP1.S1
PORTS S1,TP1_KP_lin;	Port4 = Sensor2.S2< − >TP2.S2
CONNECT TP1_WS.S1 IN VIA TP1.S1;	
CONNECT TP1_WS.TP1_KP_lin INOUT KP_WS.KP_TP1_lin VIA	
TP1.TP1_KP_lin;	
COLLECTION TP2_WS;	
MODULES	
TP2_WS_M1	
EXPORTS (Sensor2)	
TASKS	
Sensor2 (TRIGGER PORT S2,DEADLINE d_Sensor)	
PORTS S2,TP2_KP_lin;	
CONNECT TP2_WS.S2 IN VIA TP2.S2;	
CONNECT TP2_WS.TP2_KP_lin INOUT KP_WS.KP_TP2_lin VIA	
TP2.TP2_KP_lin;	
CONFEND;	
PERIPHERALS:	
NAME: Sensor1;	
INTERMESSAGE PERIOD: 10 MICROSEC;	
NAME: Sensor2;	
INTERMESSAGE PERIOD: 10 MICROSEC;	
PERIPHEND;	
ARCHEND;	

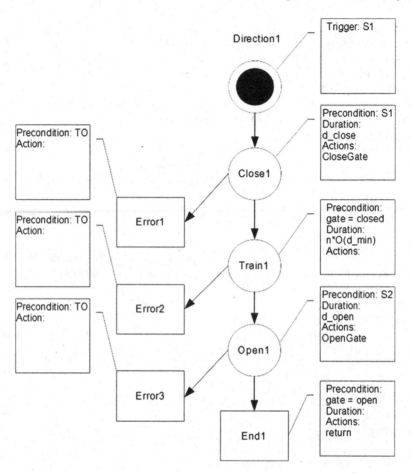

Fig. 2.5 The TSTD to program solution for task Direction1

Table 2.1 shows the textual description in Specification PEARL syntax on the left and the internal interpretation of the architecture description for the example on the right-hand side. Figure 2.5 and Table 2.2 show the TSTD diagram and the internal representation of the task handling a one-way passage of a train—from S1 to S2. The other way around is the same—only the order, in which the sensors send their signals, changes.

The internal representation of the system's architectural elements (e.g. Tables 2.1 and 2.2) is used for storing architecture data in a uniform manner for all model elements. The additional properties and TSTD code fragments are stored separately in the element database. The internal representations are used within the associated CASE environment to support the life-cycle of Specification PEARL projects—to be able to create, update, simulate and deploy the designed models. The layout of Specification PEARL projects is further described in Appendix E.

Table 2.2 Internal representation of TSTD Direction1

[State types]
START = 0
TRANSIENT = 1
END = 2
[States]
Sensor1 = START
Close1 = TRANSIENT
Train1 = TRANSIENT
Open1 = TRANSIENT
Error1 = END
Error2 = END
Error3 = END
End1 = END
[Sensor1]
Precondition = S1
MinT = 0
MaxT = 0
Next = Close1;
Action =
[Close1]
Precondition =
minT = 0
maxT = d_close
Next = Train1; Error1;
NextTO = Error1
Action = REQUEST SEMA1;
[Train1]
Precondition = NOT TRY SEMA1
minT = 0
maxT = n*d_min
Next = Open1; Error2;
NextTO = Error2
Action =
[Open1]
Precondition = S2
minT = 0
maxT = d_open
Next = End1; Error3;
NextTO = Error3
Action = RELEASE SEMA1;
[Error1]

(continued)

Table 2.2 (continued)

Precondition = TO
Action =
[Error2]
Precondition = TO
Action =
[Error3]
Precondition = TO
Action =
[End1]
Precondition = TRY SEMA1;
Action =

References

1. 66253 Part 3: Pearl for distributed systems. Technical Report, DIN (1989)
2. Colnarič, M., Verber, D., Gumzej, R., Halang, W.A.: Implementation of hard real-time embedded control systems. Real-Time Syst. **14**(3), 293–310 (1998). doi:10.1023/A:1007920407968. http://dx.doi.org/10.1023/A:1007920407968
3. Gumzej, R.: Holistic embedded control systems design with specification pearl. 1 CD-ROM. http://www.rts.uni-mb.si/misc/projekti/SPEARL/ (2006)
4. http://de.wikipedia.org/wiki/RTOS-UH
5. Pearl - process and experiment automation realtime language. http://www.pearl90.de/ (2014)
6. Gumzej, R., Lu, S.: Modeling distributed real-time applications with specification pearl. Real-Time Syst. **35**(3), 181–208 (2007)
7. 66253 Part 1: Basic pearl. Technical Report, DIN (1981)
8. 66253 Part 2: Full pearl. Technical Report, DIN (1982)

Chapter 3
Specification PEARL Methodology

3.1 System Life-Cycle

As shown in Fig. 3.1, the hardware and software configurations, as specified with the previously described Specification PEARL language, are merged into an architecture specification. Together with the application tasks, as designed with the before mentioned TSTD diagrams, they form the system model.

To deploy the system model a virtual machine is composed from all previously mentioned components. Deployment is a twofold process, since a virtual machine can be deployed to the physical environment for execution or simulation environment for verification and validation by co-simulation. Beforehand, however, the system model needs to be checked for coherency and consistency. Deployment to the simulation environment usually preceeds deployment to the physical environment to determine, if and how the anticipated QoS can be fulfilled. Through multiple iterations of fine tuning the model the expected QoS can be reached and finally, the system model is deployed to the physical environment.

To deploy our system model to the physical environment a virtual machine is cross-compiled for the target platform. For co-simulation, however, a virtual machine of the system is integrated with a simulation engine for verification and validation of its correctness, timeliness, predictability and dependability by discrete deterministic co-simulation.

3.2 System Model

The system model is composed of hardware and software models. The hardware model is represented by stations, representing the processing nodes of a system. They are characterised by their names, types and their components' properties (e.g. processor (clock frequency), memory (amount, access time), devices (interfaces, timers, etc.)). There are four different types of processing nodes in a system

© Springer International Publishing Switzerland 2016
R. Gumzej, *Engineering Safe and Secure Cyber-Physical Systems*,
Studies in Computational Intelligence 632, DOI 10.1007/978-3-319-28905-2_3

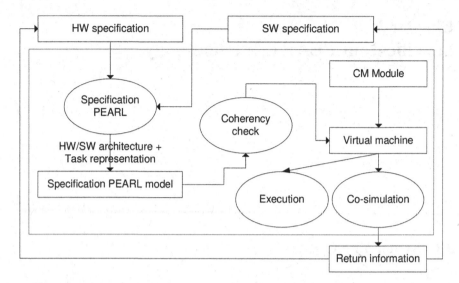

Fig. 3.1 Specification PEARL methodology

architecture: BASIC (application program and operating system), TASK (application program), KERNEL (operating system) and COMPOSITE (multi-station node), and each of them has some type-dependent properties in addition to the already mentioned general properties. A processing node may have one or more communication lines attached to it, each one connecting it to another node.

The components of the software model are collections of tasks, which are mapped to the stations of the hardware model. They are composed of sub-layers of nodes representing program tasks. The tasks themselves are represented by timed state transition diagrams (TSTD). For inter-task co-operation, collection ports are used that represent references to "physical" communication lines between stations (interfaces) of the hardware model.

3.3 Virtual Machine

The system model represents the core of the virtual machine running the designed application. To establish and run the initial configuration of collections as well as to administer configurations according to the changing station states, a configuration manager (CM) module with the optional real-time operating system is required at every station. It forms a middle layer between the hardware platform and collections of application tasks. Its role is to function as (1) a hardware abstraction layer, (2) a hardware/software interface, and (3) as an "inter-collection" co-operation agent. The attributes of a station's internal devices provide the values for the parameterisation of the station's configuration manager (CM) and real-time operating system.

3.4 Simulation Model

While being designed on separate layers, the mapping of collections to stations is made explicit for co-simulation. The model used in co-simulation is an internal representation of a system designed. The structure of the simulation units is shown in Fig. 3.2.

The structure of task collections as well as their interconnections are taken into consideration when mapping task collections' to stations' simulation units, where they are to execute. In simulation a composite station merely represents a "super-simulation" unit composed of two or more station simulation units. Hence, only their constituent nodes take part in co-simulation. Collection simulation units are linked as sub-nodes to their associated stations, whereas task simulation nodes are linked to their collections' units. For communication among tasks at different stations the appropriate collections' ports are used. A station's CM determines when a certain collection is active and dispatches its messages accordingly. tasks are represented by timed state transition diagrams (TSTD), whose program representations are used to "drive" task simulation units. During simulation their execution is responsible for advances in time and state spaces.

Co-simulation is based on the following pre-dispositions:

- there is only one global simulation clock in a system, and all STATIONs' real-time clocks (timers) relate to it (by perfect synchronisation),
- the time events relate to the corresponding station's real-time clock,

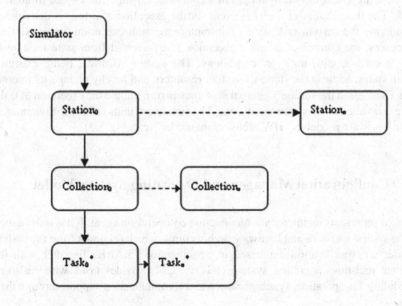

ᵗaskTSTD representation

Fig. 3.2 Structure of simulation units

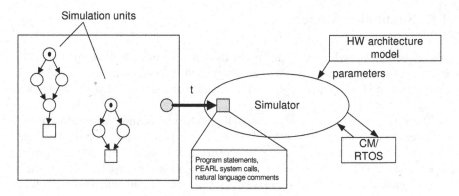

Fig. 3.3 Course of simulation

- tasks are assigned deadlines for their execution (the only exception are short initialisation tasks),
- to task states (TSTD) time frames (minimum and maximum time) for the activities performed within the states (in real-time clock time units) are assigned,
- all simulation nodes are derived from a common type of simulation unit.

During simulation, station and application task simulation units are "executing" the application—the states of tasks represent their behaviour, semantics and dynamics. The units are synchronised by means of a common real-time clock—the simulation clock. The time instant of the next event for the execution of a task is determined considering the current task state's minimum/maximum execution times. As time progresses, the control of a task's execution is transferred from state to state in correspondence with their pre-conditions. The system requests, being executed within states, address the station's system resources and hereby change its internal state. Changes in the station's internal state may in turn initiate the execution of tasks being scheduled on these pre-conditions. The simulation units refer to the parameters of the hardware model as "HW architecture model" (c.p. Fig. 3.3).

3.5 Configuration Manager and Operating System Model

The CM represents an inter-station/collection co-operation agent. It has information on the system software and hardware architectures, which originate from the system architecture specification expressed in Specification PEARL. Together with the optional real-time operating system (RTOS), CM provides tasks with real-time scheduling, co-operation, synchronisation and communication support through their APIs.

Functionally, the CM and RTOS have the same role in the co-simulation (Fig. 3.3) as in execution on a target platform. The main differences lie in the global real-time clock maintained by the simulation environment, and the handling of external signals (interrupts) as internal ones. Context switches are handled by the real-time operating system, but they are performed on a higher level. Here, the *context* refers to task states only—not processor registers.

Each processing node for a real-time operating system maintains a real-time clock. In a simulation environment, all these clocks are perfectly synchronised with the global simulation time which, in an execution environment, should be implemented by an independent global time source and a predictable time dissemination mechanism.

Pre-emption points are the same in simulation as in target platform implementation, viz task state transfers. The resource access functions and interface device drivers of a station refer to the virtual machine in case of simulation.

The time required to execute the operating system itself (schedule and dispatch cycle) is assumed constant. This time is considered to be a part of the system call service time and is, therefore, not modelled separately. The time needed to service a system call is considered to be included in the time frame of the calling task's state. Its sole function is to change the system state and to trigger task states, whose trigger conditions relate to the internal data structures of the (operating) system.

3.6 System Verification and Validation

Verification and validation of Specification PEARL models is based on co-simulation with earliest deadline first (EDF) scheduling and time boundaries. It is primarily meant to profile the timing properties of designs in order to make them feasible. A design is transformed into an internal representation for simulation, whose primary result is a successful execution or a failure, whereas the secondary result is an execution trace, from which additional profiling information is extracted. This is used to discover bottlenecks and unreachable states, as well as to fine-tune the resource parameters and to balance the load on the designed prototypes.

For successful verification, it is assumed that the designed system model is consistent. Intermediate checks on the following points may be performed during the design of the system architecture, and a final check has to be performed prior to verification to ensure this:

- Completeness check: all components are present and fully described,
- Range and compatibility check: parameter compatibility among components, and
- Software to hardware mapping check: complete coverage and consideration of resource limits.

These checks represent the preparation to verification of correctness and to validation of temporal feasibility, which is described in the forthcoming sections.

3.6.1 Verification and Validation of Temporal Feasibility

3.6.1.1 Criteria Function

Every verification method requires the definition of a criteria function, which tells, when a system fails, i.e. what the limits of the "normal" execution of the system, being checked, are. The concept of correctness had been defined as follows: "A system fails, if it reaches an undefined state during co-simulation, or if any of its pre-defined time frames is violated and no time-out action is provided."

By trying the shortest ($minT$) and taking the longest ($maxT$) transition times through every tasks' TSTD state, it is assumed that herewith a sufficient part of the time domain can be covered to allow generalising the results for each task state and, herewith, also the task as a whole to an arbitrary transition time instant within the interval ($minT, maxT$). Generalising this result to all tasks in a system renders the temporal feasibility of the system as a whole.

3.6.1.2 Co-simulation with EDF Scheduling

For verification and validation of temporal feasibility, deterministic next critical event simulation and earliest deadline first (EDF) scheduling are used. The next critical instant is always determined by the simulation unit whose activation time is the closest. This time is forwarded to all its parent units and, finally, becomes the next global critical time instant. In each step it is checked, whether timing errors have occurred. A *time-out* represents a controlled program fault, which is handled by a *time-out action* and by transition into the initial state. If such an action is not defined for the current state, the system fails. Co-simulation with EDF next-event scheduling is based on the following *timing information* (see Fig. 3.4):

- A: task activation time,
- R: accumulated task run time (updated with the next critical event),
- E: task end time (the time when the normal task end is expected based on its maximum run time; upon a context switch the current time t_1 needs to be remembered, because for re-running the task this parameter needs to be reset based on the current time t_2 and the formula $E' = E + (t_2 - t_1)$),
- D: task deadline (set, when A is known).

A task is re-scheduled when it is activated due to a scheduled event or on request. The task with the earliest deadline is chosen for execution, and its current state determines the next critical moment based on the current time t. The states of tasks are

Fig. 3.4 Task run with a single context switch (see text for abbreviations)

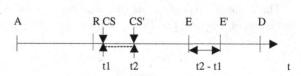

executed atomically—a context switch is not performed before a task state is worked off. A real-time operating system's scheduler is responsible for task scheduling (determining the most urgent task), while the simulator is responsible for determining the next critical moment for the current operation (task state or external event).

While re-scheduling, the following criteria (failure conditions) need to be checked for all active tasks:

- $t < Z = D - (E - (A + R))$, where Z represents the latest time when the task needs to start/continue in order to meet its deadline;
- $t < E \leq D$ must be true for all active tasks, since otherwise they have missed their deadlines.

Tasks can be scheduled to be executed upon external events. For simulation purposes, occurrence times are assigned to them. They are represented as native station unit events, whose next critical time instants are taken from an occurrence table, which lists interrupt numbers with their corresponding occurrence times. When the events occur, they are handled by the station's real-time operating system waking up appropriate tasks.

During co-simulation, the time of progression to the next state is calculated in two variants for each state:

1. $RTC + minT$ to check the pre-conditions, and
2. $RTC + maxT$ for transition to a new state.

If upon reaching the second variant of the critical time instant the pre-condition for transition to any further state is not fulfilled, the on-time-out action is executed. If the latter is not provided, the system fails. During simulation, the parameters E and D are set for each task when it is activated (the parameter A is set). When a critical instant is reached, it is checked if herewith the time-frame as given for the task has been violated, which results in the following consequences:

1. subtraction of the overhead from the task's slack time, or
2. the system fails as the task deadline is missed.

The simulation results are logged during the execution of each simulation unit, and every step is accounted for within all parent simulation units, too. This means that every task state logs its actions into the TASK-log, whereas a task logs its state changes into the COLLECTION-log. A collection logs the time when it was first allocated to a station, possible subsequent re-loads and the changes of states which triggered them into the STATION-log. The stations and collections also log the times when they were communicating among each other. All exceptions are logged where they are discovered.

3.6.1.3 Interpretation of the Results

The simulation logs are checked manually for irregularities, which could represent faults in the original design, or timing/synchronisation errors that might have occurred during the virtual "execution" of the system model.

Busy and idle times are considered for each station and, if necessary and possible, load-balancing actions are taken. The process of analysing and fine-tuning, also known as profiling process, cannot be unified due to the great diversity of possible designs. For this reason, it must be carried out manually and remains the responsibility of the designer.

The temporal feasibility, as determined by co-simulation, retains its validity if the execution times provided do not change when the software model is deployed to the physical environment. To detect possible changes, schedulability analysers can be employed, taking respective station processing capacities and program loads into consideration.

Chapter 4
UML 2 Profile for Specification PEARL

The Unified Modelling Language (UML) provides constructs to deal with varying levels of abstraction in modelling to visualise and specify both the static and dynamic aspects of systems [1]. Its notation defines the semantics of an object meta model to capture and communicate object structure and behaviour. In its meta model architecture, UML supports the extension mechanisms *stereotypes*, *tagged values* and *constraints*, which allow to tailor it towards the needs of specific domains.

A *UML profile* is a pre-defined set of extension mechanisms for a particular domain, technology or methodology, which provides a connection of how to apply and specialise UML to this domain. A *stereotype* provides a way to define virtual sub-classes of UML meta classes with additional semantics. It can set constraints additional to the ones of its base meta model class as well as tags to define further properties. A *constraint* is a semantic restriction represented as a text expression, which is usually formulated in the object constraint language (OCL). Constraints are attached to one or more model elements. Tag definitions specify new kinds of properties as part of a stereotype definition. The actual properties of individual model elements are specified using *tagged values*.

The process of defining a general UML profile for a given platform or application domain can be summarised as follows:

- First, a set of elements comprising platform or system and the relationships between them need to be defined, which can be expressed in terms of a meta model, i.e. the meta model includes the definition of the domain entities, the relationships between them and the constraints that govern both structure and behaviour of these entities.
- Once the domain meta model is built, the UML profile is defined, in which a set of stereotypes is defined for each relevant element of the meta model.
- Tagged values should be defined as attributes that appear in the meta model. They include the corresponding types and initial values. The domain restrictions are expressed by constraints.

A UML profile was built to describe the constructs and capture the essential semantic concepts of Specification PEARL. In this language, *Station* and *Coll*ection are

© Springer International Publishing Switzerland 2016
R. Gumzej, *Engineering Safe and Secure Cyber-Physical Systems*,
Studies in Computational Intelligence 632, DOI 10.1007/978-3-319-28905-2_4

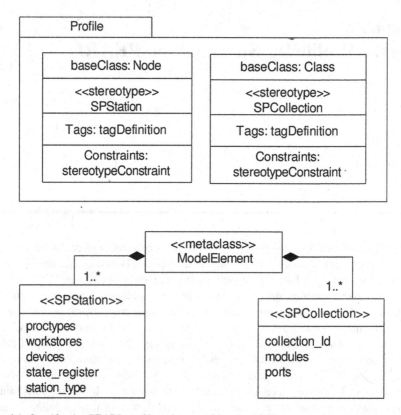

Fig. 4.1 Specification PEARL profile and compositional model

defined as basic entities in terms of the corresponding UML stereotypes, as shown in Fig. 4.1. The entities (stereotypes) will be used to define the class diagrams that specify the compositional model of Specification PEARL, i.e. the structure and relationships between the model entities. The compositional model can be used for application frameworks built with Specification PEARL elements.

4.1   Mapping Specification PEARL Architecture Constructs to UML

As described earlier, Specification PEARL includes elements to describe hardware and software configurations of distributed systems. To map these constructs onto UML elements, it is indispensable to compare UML and Specification PEARL constructs, to be able to choose appropriate base elements and define UML stereotypes for Specification PEARL elements. The essential point of this mapping is the

S-PEARL Element	UML Element	Stereotype	Icon
Station	Node	<<SPStation>>	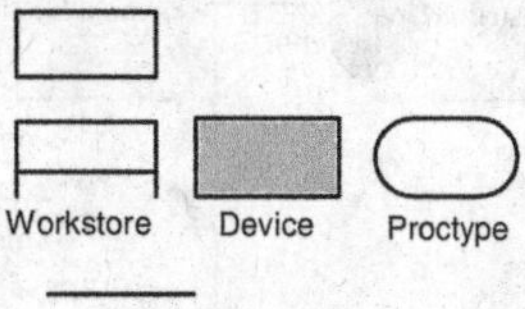
Component	Class	<<SPComponent>>	Workstore Device Proctype
Line	Connector	<<SPLine>>	
Collection	Class	<<SPCollection>>	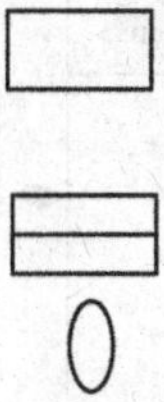
Port	Class	<<SPPort>>	
Module	Class	<<SPModule>>	
Task	Class	<<SPTask>>	

Fig. 4.2 UML stereotypes for Specification PEARL (S-PEARL) constructs

Specification PEARL architecture, its real-time features and its run-time constraints. Figure 4.2 shows the main stereotypes defined for Specification PEARL.

4.1.1 Station Layer

In Specification PEARL, hardware and its deployment is introduced on the station layer. The processing nodes (stations) of a system are treated as black boxes with connections for information exchange. On their subordinate layers, stations are described by the properties of their components, such as *Proctypes*, *Workstores* or *Devices*. There may be many stations in a system, each one being uniquely identified, and equipped with an abstract *state register variable* for re-configuration purposes. Stations communicate among each other through the connections established, which are defined on the component layer by hardware devices of type *interface* being referenced through *ports* of the software architecture.

A node in UML is a run-time physical element that represents a computational resource, which may be instantiated and stereotyped to be distinguished between different kinds of resources. Associations among nodes represent their communication paths. They can be stereotyped to distinguish between different (types of) paths. Nodes have unique names. They may hold objects and component instances and represent the physical deployment of components. Therefore, it is natural to describe stations and net-connections of Specification PEARL in UML by nodes and their associations, and to define corresponding stereotypes for various types

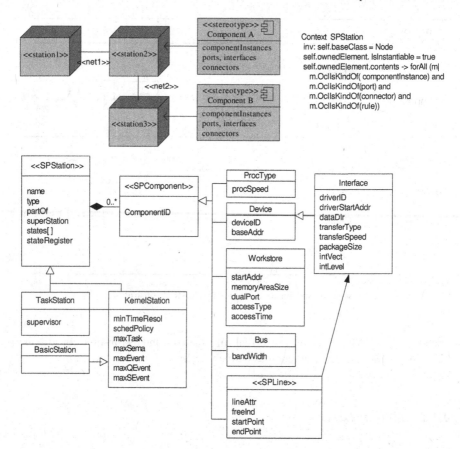

Fig. 4.3 Station stereotype (deployment diagram) and structure of the hardware part of "Architecture Data" (class diagram)

thereof. Figure 4.3 depicts a station in Specification PEARL and defines it in terms of UML deployment and class diagrams.

4.1.2 Collection Layer

In Specification PEARL, collections are introduced as the largest separately loadable software components, to which states are assigned while they are active on the stations they are loaded to. Collections are composed of modules of tasks. Communication between collections is performed on the basis of the port concept by message exchange, only. When a station changes its state, another collection is activated and the new connections are established. The collections loaded to the same station are grouped into a *configuration*. They are administered by the *configuration manager*

(CM), which chooses the active collection and dispatches messages among collections (also at different stations) through their ports.

In UML, a *component* can be a modular, replaceable and deployable piece of software that is available at specification time, at deployment time and at run time. A component's internal structure also shows how it interacts with its environment— exclusively through interfaces or, more often, through ports. Therefore, a component can be replaced by another one which offers at least the same provided and required interfaces or ports, as these are the only parts of a component which are accessible by its environment. Physical instances of software components can be deployed on nodes.

The components of UML are composed of *parts, connectors, ports* and *interfaces*. They exchange data with each other through ports. Viewed from the outside, a component is a set of provided and required interfaces, which may be exposed via ports. Internally, it is a set of class instances or parts that collaborate to implement the services exposed by the component's interfaces. Parts represent sub-components. In Fig. 4.4 a component meta model is defined as an extension of UML components by adding the non-functional aspects contract and general properties. This figure illustrates the component concepts and reflects both external and internal views. A component owns a unique identifier and a set of properties, and defines a set of communication ports which provide interfaces. Components can exchange data with each other through ports and connectors, only. A component may be *composite*— containing other component(s).

With its behaviour and elements (modules and ports) as shown in Fig. 4.5, the configuration of collections in Specification PEARL shares greatest similarity with a component in UML, as both of them represent primary computational elements,

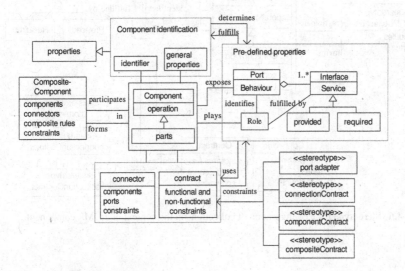

Fig. 4.4 UML component meta model

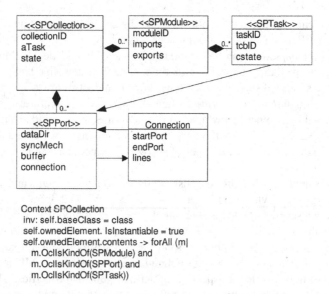

Fig. 4.5 Structure of the software part of "Architecture Data" with stereotypes for collection, module, task and port

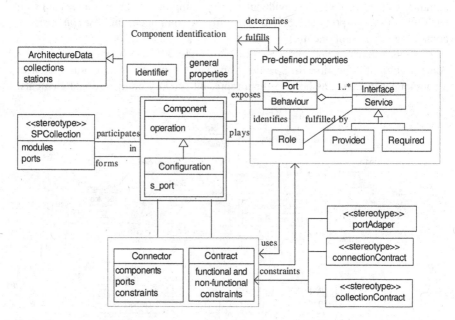

Fig. 4.6 Stereotype for a configuration of collections in the context of a UML component

both have ports, both may be decomposed hierarchically and both are replaceable. Thus, it is natural to associate a configuration of collections with a component (see Fig. 4.6), and a collection, as its part, with a class as shown in Fig. 4.5:

Connector   is a link in the component meta model that may be of kind delegation or assembly. A delegation connector either links a provided port of a component to a part of the component's realisation, signifying that requests, received through the port, are forwarded to the part, or it links a realisation part to a required port, signifying that requests sent through the port originate in the part. Several connections may exist between a single port and different realisation parts. An assembly connector links a required interface or port of a component to a matching provided interface or port of another component.

A connection in Specification PEARL represents a link between ports of collections. For the same purpose in UML, the connector is used to link components, or sub-components through port-to-port connections. Thus, connections can be mapped to UML connectors.

In Specification PEARL, collections communicate among each other by port-to-port message exchange, which avoids direct references to communication objects in other collections, and decouples the communication infrastructure from the logic of message passing. One-to-many and many-to-one communication structures are allowed. A message may be sent using either an asynchronous "no-wait-send", a synchronous "blocking-send" or a synchronous "send-reply" protocol. Synchronous sends and receives may be specified with time-out clauses. The main purpose of protocols in Specification PEARL is the definition of communication patterns, i.e. patterns of messages sent from one collection to another. In UML, protocols represent the behavioural aspects of connectors, which are similar to the communication patterns in Specification PEARL. Thus, we can define constraints and tagged values for the communication patterns, and assign them to ports and connections in order to achieve similar effects as in Specification PEARL.

Port   is a named and typed interaction point of a component in the component meta model. A provided port is characterised by a provided interface, a required port by a required interface, and a complex port by an arbitrary set of provided and required interfaces. Complex ports enable the localisation of complex interaction patterns where calls may occur in both directions. Unlike interfaces, a port may be associated with a behaviour, specifying the externally observable behaviour of the component when interacting through the port. This allows the specification of semantic contracts. A component may have multiple ports typed by the same interface, and is able to distinguish between calls received through different ports. In Specification PEARL, there are in-, out- and in-out ports which could directly be mapped to ports in the component meta model, since both serve as interfaces that define points of interaction between the computational elements and their environments. We have defined, however, a Port stereotype for inter-collection communication with Specification PEARL port properties and functionality. We use a dedicated component port "s_port" to transfer system call parameters in asymmetrical systems which are serviced through the CM object (see Fig. 4.8).

Interface   is the only part of a component visible to users. It should provide all the
        information that the users need in order to deploy the component, and contain
        specifications for its operations. It is a set of operations that is used to specify a
        service of a class or a component. During execution they are used when invoking
        the component's functionality by the application.

    In   Specification PEARL, the collections are called through uniform interfaces,
        and their only points of interaction are the ports mentioned before. Their exe-
        cution and collaboration is organised by the configuration manager being the
        primary execution class of any component at any station.

Properties   are used to characterise aspects of components. General properties can
        be expressed with respect to timing and resource usage such as deadline, time
        period and worst-case execution time (WCET), or resource consumption. Pre-
        defined properties are used to express super-component, ports or constraints.

    Timing   requirements could be expressed as *TaggedValues* attached to the
        "Task" stereotype of a "Collection". The task stereotype can, however, also
        hold this information. When several tasks are ready to run, a priority-driven
        scheduler should select the task with the highest priority to execute. Sched-
        ulers are timed systems that manage shared resources. Usually, schedulers
        apply scheduling policies to select among pending requests to allow for access
        to resources. The scheduler polices can be expressed by contract.
    Similar   properties also pertain to other Specification PEARL constructs and
        may be assigned to them as properties. The assigned properties are meant for
        system programs, which have to know how to interpret them. Hence, these
        features are dependent on the target platform and have to be used with caution.

Contract   is defined as a class in the component meta model used to specify a
        component's operation constraints. To specify functionality it uses the theory
        and methods of the design-by-contract approach [2]. It can be assigned to a
        port, connector or component, and govern some functional or non-functional
        constraints. Furthermore, architecture constraints can be divided into ones for
        components, composition and connections. In real-time systems, a component
        constraint may describe a property of temporal criticality, which its environment
        expects from a component. A connection constraint describes time criticality
        of message transmission across components which is, normally, a system-wide
        (or subsystem-wide) timing requirement. A composition constraint describes the
        time behaviour expected by a component from its environment. Also, a UML
        operation contract can be employed that identifies system state changes when an
        operation takes place. Effectively, it will define what each system operation does.
        All constraints can be specified by employing contracts and assigning them to
        corresponding participants.

Operation   specifies an individual action that a component object will perform. It
        deals with input parameters which specify the information provided or passed
        to the component, output parameters which specify the information updated or

returned by the component, any resulting change of the component's state, and any constraints that apply.

Port adapter   enables the connection of two incompatible ports. It defines the semantics associated with the ports and provides the operations, which are expected from the respective other port. The adaptation is realised at mapping time. A port adapter can also describe time-dependent, operational-behaviour constraints of components.

Composition components   specify how components are interconnected. They contain a number of component instances and define their configurations. In addition, a composition component also specifies how the ports of those instances are wired, i.e. which connector is used to connect which ports. It is defined for the purpose of configuration, and may occur in parts of components or in a main component such as system composition. A composition component contains a number of connected sub components, rules which specify compositional constraints, and component ports, which may form internal ports of the composite. A composite component also has external ports, which are the only ones visible from the outside. The external ports are connected to appropriate internal ports and connectors.

4.1.3 Binding the Specification PEARL TSTD to UML's State Chart Concept

For proper task representation and management some additional constructs still need to be defined for use in UML models. In UML, state machines are adopted to model the dynamic aspects of a system, which focus on the event-order behaviour of an object and show the event-triggered flow of control due to transitions leading from state-to-state. A state machine models the lifetime of a single object, whether it is an instance of a class, a use case or even an entire system. An object may receive an event, respond with an action, then change its state, and it may also receive another event. Its response may be different, depending on its current state in response to the previous event.

A state chart representation is chosen to model adaptive operational behaviour. The modelling objects provided are states, events and transitions: (1) states represent the operational model, (2) events represent the causes of mode shifts and (3) transitions and transition rules define the pre-conditions and the consequences of mode changes. States can contain states, events and transitions, thus enabling the creation of hierarchical finite state machines. Therefore, the UML state chart formalism can be used to model Specification PEARL's task concept by defining a translation to the task's "main()" method.

In Fig. 4.7, the UML state chart mechanism is shown. It includes *CM, Event, ActiveObject, StateTransitionTable, State, Transition* and *Activity*. The necessary adjustments for implementing Specification PEARL timed state-transition diagrams (TSTD) are discussed below:

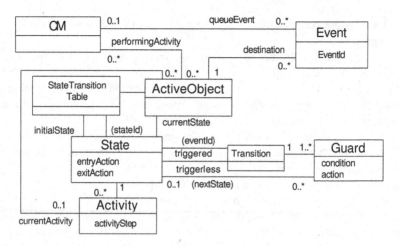

Fig. 4.7 CM in the context of UML's state chart mechanism

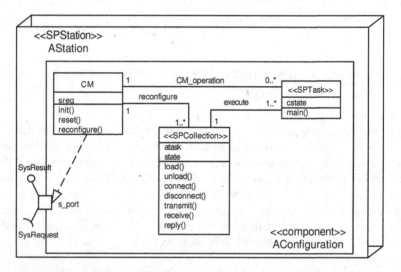

Fig. 4.8 Specification PEARL-oriented application architecture in UML

- The main flow of control is based on events rather than function calls. The *CM* object, as local executive for each station, controls the execution path at this (part of the) system (see Fig. 4.8). It also keeps a list of active objects (e.g. collections/tasks in Specification PEARL) that are currently executing activities. Switching between dispatching events and executing activities allows the other active objects in the system to process, and also allows the activities to be interrupted by incoming events.

- A *state* reflects a situation in the life of an object during which this object satisfies some condition, performs some activity or waits for some event. An object remains in a state for a finite amount of time.
- A *transition* indicates a change from one state to another, indicating that an object in the first state will perform certain actions, and enter the second state when a specified event occurs and other specified conditions are satisfied. Each transition has a label that comes in three parts [3]: *trigger signature, guard* and *activity*. All parts are optional. The trigger signature is usually a single event that triggers a potential change of state. The guard, if present, is a Boolean condition that must be true for the transition to be taken. The activity is some behaviour that is executed during the transition. It may be any behavioural expression (e.g. program statements or PEARL system calls). The full form or a trigger signature may include multiple events and parameters.
- An *event* class defines the functions that dispatch an event to its destination active object. Each event carries the identifier of the active object that will receive the event.
- The *StateTransitionTable* consists of a set of states defined by *ActiveObject*. It also maintains the *initial state* to enter when a new instance of *ActiveObject* is created.

In the Specification PEARL methodology, an executable program is a collection of modules, being composed of a set of tasks that respond to events (see Fig. 4.5). Tasks represent the processes of a running system, i.e. *active objects* in the UML model. They are modelled in Specification PEARL by TSTD diagrams. The translation of TSTD diagrams to task prototypes relies on state enumeration, which enables the execution- and collaboration-managing *CM* to switch among active tasks and states, and to return to the previous state upon resumption of the previously active task. Pre-emption points concur with task state transfers (i.e. a context switch shall occur when a task state is worked off) as modelled in its source TSTD diagram. If the mentioned enumeration is introduced in the translation of a UML state chart to task prototypes, these two formalisms may be used interchangeably. The states in TSTD diagrams can be assigned time/event trigger conditions and minimum/maximum times for their execution. These parameters are taken into account during the translation to task prototypes (e.g. Fig. 2.3) for the generation of appropriate system calls to the CM's inherent real-time operating system.

Since the translation of UML state charts depends on tools (and target platforms), the system calls and timing limitations should be coded into state chart actions and CM steering actions. In order to be interpreted correctly, however, *CM* and *architecture data* libraries need to be combined in the final compiled project.

4.2   UML Application Architecture with Specification PEARL Stereotypes

An application architecture should consist of a set of station nodes and configuration components, and a set of static or dynamic links that may be established during the application's execution. As shown in Fig. 4.8, the CM is described as a global configuration object that performs the run-time re-configuration of stations and collections according to the application architecture.

In this application architecture, the *ArchitectureData* package is defined as part of the configuration. It stores the relevant information about the system architecture which forms (a part of) the application. It also specifies dependencies that exist between station stereotypes and UML nodes for deployment. This information is represented in the UML model by the parameterised stereotype objects *SPStation* and *SPCollection*, respectively, representing *ArchitectureData*, whose structure is outlined in Figs. 4.1, 4.3 and 4.5.

The configuration manager (CM), being the base class of each station's *Configuration* component, is responsible for *Collection* activation and deactivation, as well as to connect and disconnect their logical communication paths, based on the stations' states.

References

1. OMG: Unified modeling language (uml) resource page. http://www.uml.org/ (2015)
2. Meyer, B.: Applying "design by contract". Computer **25**(10), 40–51 (1992). doi:10.1109/2. 161279. http://dx.doi.org/10.1109/2.161279
3. Fowler, M.: UML Distilled: A Brief Guide to the Standard Object Modeling Language, 3rd edn. Addison-Wesley Longman Publishing Co. Inc, Boston (2003)

Chapter 5
UML Safety Pattern for Specification PEARL

A safety pattern was defined based on the re-configuration management pattern, and the architectural specifications in Specification PEARL. It is meant to be used for real-time applications to be developed with UML-RT.

During re-configuration, application data must remain consistent and real-time constraints must be satisfied. In order to be able to achieve this, these issues must be addressed at multiple levels of CPS design. At the lowest level, the hardware must be re-configurable. Software-programmable hardware components support this inherently, since their functions can be changed by their memory contents. Internal hardware structures are designed to restrict dangerous conditions that could damage hardware. At the next higher level, the internal states of the software must be managed under changing tasking. Operating systems support flexible implementations of multiple tasks on single processors in form of time-sharing and/or multitasking. On the top level, one wants to define operation scenarios—configurations—for an application, which enable it to adapt to varying conditions in the environment on one hand, and to respond to changing operational modes by switching between operation scenarios in a safe and predictable manner on the other. Typically, these configurations cannot be managed by operating systems, since groups of processes and possibly also hardware components are involved. Hence, their management is usually placed on the application or middleware level, since it requires the observation of and actions based on the system state. Generally, by this approach, low-level efficiency and hard real-time properties are difficult to achieve. Because of this, the decision was to distribute re-configuration management to all three levels—hardware, middleware and software. With this in mind, the hardware/software co-design profile and pattern for real-time application design in UML based on the specification language Specification PEARL were developed. While in the profile the constructs of Specification PEARL are introduced with their properties, behaviour and interconnections, the configuration management pattern provides the mapping of software-to-hardware components, and a foundation on which to build custom CPS applications. This approach is followed in extending and parameterising the configuration management pattern with safety and security features. The pattern and

© Springer International Publishing Switzerland 2016   53
R. Gumzej, *Engineering Safe and Secure Cyber-Physical Systems*,
Studies in Computational Intelligence 632, DOI 10.1007/978-3-319-28905-2_5

its safety-oriented use are presented throughout this chapter with the goal to construct a safety shell (cf. [1, 2]) for an application designed.

In this chapter, the implementation of the safety shell features as defined in [2], namely its timing and state guards as well as I/O protection and exception handling mechanisms, is presented. The pattern is parameterised by defining the properties of its components as well as by defining the mapping between software and hardware architectures. Initial and alternative execution scenarios as well as the method for switching between them are defined. The goal pursued with the safety shell is to obtain clearly specified operation scenarios with well-defined transitions between them. To achieve safe and timely operation, the pattern must provide safety shell mechanisms for an application designed, i.e. enable its predictable deterministic and temporally predictable operation now and in the future.

In terms of safety, first, the methods are addressed that should assure correct service of a system during its entire up-time. They are meant to minimise the possibility and provide appropriate handling of system failures. While a failure represents a transition from correct to incorrect service (i.e. to not implementing the system function), the methods and mechanisms of the safety shell are intended to minimise the possibility of failures, provide service restoration, as well as minimise the time of service outages. Moreover, faults should be contained and errors handled in real time with respect to application deadlines.

With the help of the safety shell features of the Specification PEARL configuration management pattern presented, distributed real-time application programs, designed with UML-RT, can run with safe, predictable behaviour and re-configuration support. Besides the application structure, the configuration management pattern also defines uniform interfaces and protocols for intra- and inter-component/-node communication using pre-defined port/interface definitions. In (hard) real-time systems, it shall provide the necessary support for deterministic and dependable dynamic system re-configuration. The safety shell features are enabled by the pattern, although selecting and using the mentioned mechanisms remain the responsibility of a real-time application's designer, since no two safety-critical real-time applications are equal.

5.1 Design for Safety

Predictability and *dependability* are major pre-conditions for reliance to be justifiably placed on CPS and their applications. Hence, in order to address safety in a broader sense, these properties ought to be considered throughout the entire life-cycle of an CPS application—from design via implementation to upgrades and maintenance. Therefore, a design pattern that would enable addressing most safety issues and build safe and persistent applications was proposed by Kornecky and Zalewski [2]. They described a "Safety Shell" for real-time applications to be composed of several "guards", each one protecting a certain part of an application providing it with safety as well as security features. Thus, the input/output would need to be protected from tampering as well as by range checking to sustain the environmental parameters of applications. Then, exception handling mechanisms should protect applications from malicious consequences of unforeseen situations, by offering mechanisms that bring them back to normal operation. Finally, application operation should be monitored and safeguarded in its state and time spaces in order to prevent applications from leaving their specified execution and temporal frameworks. Since all mentioned mechanisms foresee different scenarios for phases of initialisation, normal operation and of various exception modes, enabling dynamic re-configurations on the application level is crucial to enable these features.

In the design and development of *dependable* CPS, the management of *dynamic (re-) configuration* has systematically been addressed by hardware/software co-design methodologists (cf., e.g. [3–5]) and the Specification PEARL methodology. Besides defining diverse (dynamic) operation scenarios, two main goals were targeted by this approach:

1. achieving fault tolerance by system design (cf. [6]) and
2. fast scenario switching (e.g. in supervisory control and data acquisition and process control systems, cf. [4, 7–9]).

The foremost property for CPS is *predictability*. It also pertains to behavioural predictability, which is addressed in the following sections. In the real-time domain, however, usually a system's timeliness is meant, which represents the property of the system whether or not all its actions are performed in time during its entire up-time. Such a system is considered to behave in a (temporally) predictable way, and is said to "operate in real time". *Dependable* CPS sustain their predictability during their entire life-time.

During design, predictability is supported by carefully planning the order of activities as well as taking care of their durations. Often the activities are executed periodically and, hence, a main loop, named *cyclic executive*, is introduced to cyclically call them up. It invokes the activities in a certain order when a timer, to which its period is assigned, times out. Naturally, the sum of their durations shall not exceed this period for all of them to finish in time. These activities, usually called *tasks*, do not necessarily have unique and fixed periods. Hence, dynamic scheduling algorithms were devised (e.g. the rate monotonic (RM) one), which dynamically assign priorities to tasks based on their relative urgencies (reciprocal values of their periods). Some or all of the tasks may not be periodic, however, but still have temporal constraints on their execution. For these cases other scheduling algorithms, not oriented at priorities, were devised (e.g. Earliest Deadline First (EDF) or Least Laxity First (LLF)). They are considered in a special kind of performance analysis, named *schedulability analysis*, which determines the temporal predictability of diverse execution scenarios to discover bottlenecks and, foremost, to provide for the temporal predictability of a system's activities.

There are two kinds of design approaches which enable reasoning on the (temporal) predictability and dependability of task executions—formal and non-formal ones. The formal design methods encompass temporal state automata (e.g. communicating shared resources (CSR) [10], timed Petri nets [11] or UML state charts). The non-formal design methods encompass Gantt diagrams and derivatives thereof (e.g. UML sequence diagrams and UML timing diagrams). While formal design methods can estimate the execution times of activities, it is still up to the scheduling algorithm to ensure their timely execution in correspondence with their periods/deadlines. In UML-RT, as in Specification PEARL, state charts are used.

Permanent readiness in high-availability systems requires that they are designed for non-stop dependable operation. Of course, also such systems require maintenance and occasional software upgrades. For this purpose, re-configuration management mechanisms were developed. Initially, static re-configuration was used, representing strictly defined operation scenarios (e.g. manufacturing lines, space shuttle or avionics). With the advent of reactive systems the need for dynamic re-configuration arose, where the links between phases are not so strict, and where there may be scenario parts that are interchanged leaving the rest intact and functioning.

Execution environments supporting dynamic re-configuration encompass the following features:

- specification of hardware and software configurations with well defined re-configuration scenarios—conditions and methods for re-allocation of hardware/ software components and their interconnections,
- local state change monitoring and delegation of state changes to affected processing nodes during re-configuration, and
- predictable overhead of minimum size to overall execution time by short and well defined re-configuration actions.

5.2   Safety Shell

The Safety Shell pattern was constructed by combining a set of UML-RT [12] stereotypes [13], which represent Specification PEARL constructs (see Fig. 5.1) as a coherent whole. They constitute building blocks at the three levels of architectural modelling, viz., hardware architecture, software architecture and software-to-hardware mapping. The hardware architecture is composed of processing nodes, termed "stations", whose descriptions also contain the properties of their components. A software architecture is organised in form of *collections* of *modules*, comprising program *tasks*, functions and procedures of the application software. The collections are units of software to be mapped onto a station. At any time there is exactly one collection assigned to run on a station. Thus, the collection is also the unit of *dynamic re-configuration*. Each Specification PEARL model is composed of stations and collections, having their specific attributes, which pertain to all objects of this type (such as properties, relations and initialisation). They are layered on three levels of abstraction (see Fig. 5.2):

1. station level, where a mapping of collection configurations to stations is established,

S-PEARL Element	UML Element	Stereotype	Icon
Station	Capsule	<<SPStation>>	
Workstore,...	Class	<<SPWorkstore>>,...	Workstore Device Proctype
Line	Class	<<SPLine>>	
Collection	Capsule	<<SPCollection>>	
Port	Class	<<SPPort>>	
Module	Package		
Task	Capsule	<<SPTask>>	

Fig. 5.1 Specification PEARL constructs and their UML (-RT) stereotypes

2. configuration level, where collections grouped into scenarios, named "configurations", are managed by a "re-configuration manager", and
3. collection level, where the tasks, which may be grouped into modules (UML packages), are managed by their collections according to their scheduling parameters.

Each collection belongs to a *configuration* and is mapped to a station. *Configuration management* is responsible for the co-operation among collections and possible dynamical re-configurations, which depend on the state changes of the stations they are residing on. A detailed description of its safety-oriented use is presented in the sequel.

5.3 Safety Shell Functionality

A safety shell is responsible for guarding the main process termed *Primary control* (see Fig. 5.3) by providing it with additional functionality which keeps the possible sources of error to a minimum. In order to be effective, these functions have to be integrated into or built around an application. In our case, the second variant was chosen by implementing a pattern, which forms the "backbone" of an application, requiring it to be formed in a specific manner in order to function in the safety shell's environment. The configuration management pattern has the structure and functions needed to fulfil the role of a safety shell in terms of guarding a system in such a way that it always remains in a foreseen state and time-frame of operation. In the sequel,

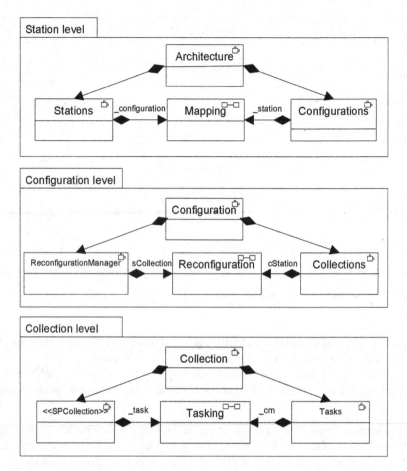

Fig. 5.2 Levels of configuration management

the functions of the four protective mechanisms are described, and it is explained in which manner and to which degree they safeguard an application's execution.

5.3.1 Protected Input/Output

Protected input/output refers to well defined interfaces with the environment. By well defined, we mean stable physical connections and sound protocols with integrated error checking and correction techniques. Usually, the possible problems originate from data overruns or malicious data. By themselves, the device drivers of interfaces can only correct a part of these problems associated with data formats and protocols.

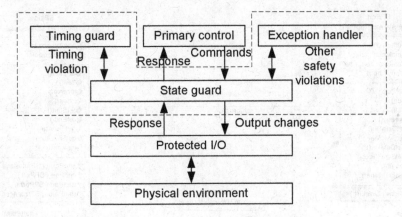

Fig. 5.3 Safety shell scheme

They could, however, also detect overruns or out-of-scope data, prevent recurring corruption of data, and signal possible errors.

In our implementation of *I/O connections* all relevant properties of communication channels are visible at the application level—data direction (IN, OUT, INOUT), data organisation (e.g. transfer unit, packet size), synchronisation method of the smallest transfer unit and number of lines used. Even with multiple lines for a connection, any connection is considered as a *port-to-port logical connection,* whilst its provided attributes steer the underlying protocol. *Ports* are application level access points to the connections—applications transmit and receive data through them.

In our implementation of *I/O ports* (see Fig. 5.4), also an important safety-related feature is present, namely routing parameters. Where stable line connections are of utmost importance, they are usually designed redundantly, being doubled, tripled or with one of the *communication lines* representing a slower yet reliable (e.g. wireless) connection. In our routing parameters, we can determine the lines which can/must be used, and/or assign a preferred line, being the fastest or most trusted one. If a line is not or becomes unavailable, the protocol automatically searches for the next available line. On the application level, it is important to have uniform interfaces between software components possibly executing at different processing nodes, which is also achieved by *port-to-port communication.* The lower levels of communication are suppressed on the application level, however, as pointed out above, the parameterisation of the connection lines between ports and device drivers is made transparent by the Specification PEARL profile, and enables complete oversight down to the physical level.

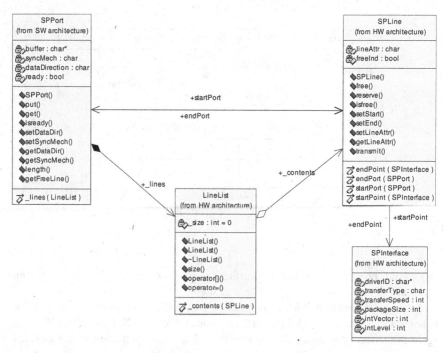

Fig. 5.4 Safeguarded port-to-port communications

5.3.2   State Guard

There should be a pre-defined scenario for each possible state a system may assume ("state guard").

Problem decomposition enables to consider loosely coupled interdependent processing nodes, which ensure local predictability, and have well defined global interconnections that form an integral part of each scenario. Due to a possible state explosion during execution, it is impossible to (pre-) determine all global states a system can assume and define scenarios for them. Since scenarios are defined for each station separately (e.g. see Fig. 5.5), this problem becomes easier to tackle. There exists a limited number of states, only, and as global interconnections are (re-) connected during re-configuration, the global implications are unproblematic. Hence, besides *fault containment*, this makes designing distributed real-time applications easier, and the systems designed more robust. A rigorously designed application structure, in which the execution of a collection of tasks is associated with an exactly predetermined state, and a simple and well defined mechanism of changing station states prevent the transition to an undefined state, herewith implementing the state guard function. The same holds for tasks, where an exactly defined activity structure does not only prevent transitions to an undefined state by exception handling, but also supports safeguarded execution of temporally sensitive operations (see Fig. 5.6).

```
// "initial_load_statement"
void init() {
    reconfigure(0); // '0' is the initial state
}

// "reconfiguration_statement"
void reconfigure(char s) {
    if (sreg!=s) { // sreg is the station state register
        switch (sreg) {
            case -1: // initially it is undefined
                break;
            default: // upon state change the current collection is unloaded
                sCollection.unload().sendAt(sreg);
                break;
        }
        switch (s) {
            case -1: // if there is no valid state, nothing is loaded
                break;
            default: // the collection associated with the new state is loaded
                sCollection.load().sendAt(s);
                break;
        }
    }
    sreg=s;
}
```

Fig. 5.5 Safeguarded state transitions within a station (configuration)

In Fig. 5.5 state-dependent switching among collections of tasks is presented. Every state switch operation consists of stopping (unloading) the currently active collection and starting (loading) another one. In addition to terminating its scheduled tasks, also the existing port-to-port connections of the active collection need to be disconnected. When the new collection is loaded—prior to initiating its initial task— its connections need to be (re-) connected. Using this approach, *fault prediction* is already included in system design, and system *availability* can be guaranteed. Due to local station state management and state propagation, at one hand fault containment is assured, while on the other smooth transitions between states are enabled also on the system level, since the global implications of local state changes are defined for every station with equal rigour.

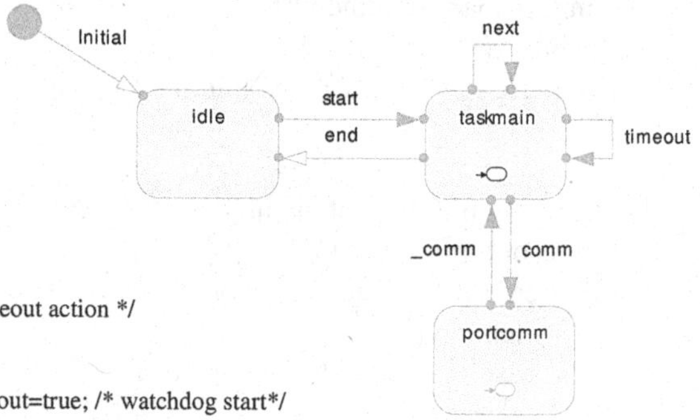

```
void taskmain()
{
      if (_start_state) {
            aState=choose_start();
            _start_state=false;
            _timeout=false;
      }
      switch (aState) {
            case i^th_state: {
                  if (_timeout) {
                        /* timeout action */
                  }
                  else {
                        _timeout=true; /* watchdog start*/
                        maxT.informIn(RTTimespec(timeout,0));
                        /* activity */
                        aState= i^th_next();
                        maxT.cancelTimer();
                        _timeout=false; /* watchdog end*/
                  }
                  break;
            }
            case end_state: {
                  /* activity*/
                  _start_state=true;
                  _timeout=false;
                  break;
            }
            otherwise:
                  break;
      }
}
```

Fig. 5.6 Temporally safeguarded execution of task activities

5.3.3 *Timing Guard*

Timeliness, being a critical property of CPS, is of utmost importance for their
applications and, hence, it is vital that their "backbone" ("safety shell") does not introduce any significant delays into execution. Due to this and to ensure observability, the
service algorithms of the pattern are kept simple. They introduce no unbounded delays
into scenario switching. This was one of the predispositions while designing them,
and is as important for safety as for timeliness of operation. Since some operations,
such as scheduling, message transmission or task activation, still require some time,
however, the operating system overhead shall not introduce any unbounded delays
either and, moreover, the service times of operating system calls have to be incorporated into task/operation execution times. Hence, the execution of an underlying
real-time operating system has also to be temporally predictable in order to enable
timeliness. In our implementation this is accomplished by a small custom real-time
operating system. Temporal monitoring of atomic activities such as executions of task
operations is possible by introducing timers into them and, in this sense, prohibit

any unreasonably long executions of atomic activities (e.g. by using a watchdog mechanism—"time guard"—see Fig. 5.6). Here, it is also possible (for any activity—task state) to define a time-out action, which is executed in response to a possible time-out condition.

5.3.4 Exception Handler

Since tasks can be re-scheduled at preemption points, only, and tasking operations are defined for active tasks, it is sensible to limit their duration. In case of violating an execution time-frame, a pre-defined scenario could be activated representing, for instance, *graceful degradation*. To further support *fault-tolerant operation*, however, it is desirable to check the correctness of other vital operation parameters, too, and by doing so implementing other features of the—"exception handler"—as well. This may introduce additional overhead, but as long as the execution times remain predictable and within the time-frames foreseen, this is not a problem. Exception handling is, in part, already present in the safeguarded I/O operations. As already mentioned, the port-to-port communication protocol also enables line replication, thus allowing for "no single point of failure" planning. Range and other error checking mechanisms could be implemented to ensure *fail-safe operation* using well known mechanisms in the same manner as we implemented the time-out handling by introducing, e.g. pre-/post-condition checking of the (critical) tasks' activities and implementing appropriate error handling routines. To achieve this, the proper application program tasks would then not only have to be designed according to the (re-) configuration management pattern, but they would also need to include appropriate built-in-test or built-in-self-test mechanisms, e.g. in the form of appropriate fault handling and of monitoring initial and/or continuous task states and other relevant process variables. The selection and appropriate implementation of these mechanisms is, however, application-dependant and, hence, exceeds the scope of this chapter.

To support *reversion modes*, several collections with the same functionality may be defined within the same configuration to be activated depending on the different operational modes. The "context" of a configuration can be maintained as a list of "collection contexts" or—as in our case—in the form of procedures to (re-) establish collections. They could be (re-) established while (re-) loading and (re-) connecting their ports when desired/needed. The choice to (re-) start or continue a collection's execution depends on the nature of the application and is, hence, left to the designer. Although one would typically continue from a state-switching condition, it is not always desirable or even dangerous to do so. In most cases, a collection is only re-loaded if its initial pre-conditions and environment state have been re-established. The "collection context" itself would consist of its task control block (TCB, i.e. task context) table as well as the lists of currently active tasks and ports. All of these would need to be re-established when re-loading a collection to continue its execution.

References

1. van Katwijk, J., Toetenel, H., Abd, Anderson, E., Zalewski, J.: Specification and verification of a safety shell with statecharts and extended timed graphs, pp. 37–52. http://dx.doi.org/10.1007/3-540-40891-6_4 (2000)
2. Kornecki, A.J., Zalewski, J.: Software development for real-time safety-critical applications. In: SEW Tutorial Notes, pp. 1–95. IEEE Computer Society. http://dblp.uni-trier.de/db/conf/sew/sew2005t.html#KorneckiZ05a (2005)
3. Kramer, J., Magee, J.: Dynamic configuration for distributed systems. IEEE Trans. Softw. Eng. **11**(4), 424–436 (1985)
4. Rust, C., Stappert, F., Bernhardi-Grisson, R.: Petri net based design of reconfigurable embedded. In: Kleinjohann, B., Kim, K.H., Kleinjohann, L., Rettberg, A. (eds.) DIPES, IFIP Conference Proceedings, vol. 219, pp. 41–50. Kluwer (2002)
5. Wolf, W.: A decade of hardware/software codesign. Computer **36**(4), 38–43 (2003). doi:10.1109/MC.2003.1193227, http://dx.doi.org/10.1109/MC.2003.1193227
6. Kalbarczyk, Z., Iyer, R.K., Bagchi, S., Whisnant, K.: Chameleon: a software infrastructure for adaptive fault tolerance. IEEE Trans. Parallel Distrib. Syst. **10**(6), 560–579 (1999). http://dblp.uni-trier.de/db/journals/tpds/tpds10.html#KalbarczykIBW99
7. Eisenring, M., Platzner, M., Thiele, L.: Communication synthesis for reconfigurable embedded systems. In: Lysaght, P., Irvine, J., Hartenstein, R.W. (eds.) FPL, Lecture Notes in Computer Science, vol. 1673, pp. 205–214. Springer (1999). http://dblp.uni-trier.de/db/conf/fpl/fpl1999.html#EisenringPT99
8. Hutchings, B.L., Wirthlin, M.J.: Implementation approaches for reconfigurable logic applications. In: Moore, W., Luk, W. (eds.) Field-Programmable Logic and Applications, Lecture Notes in Computer Science, vol. 975, pp. 419–428. Springer, Berlin (1995). http://dx.doi.org/10.1007/3-540-60294-1_136
9. Jean, J., Tomko, K., Yavgal, V., Cook, R., Shah, J.: Dynamic reconfiguration to support concurrent applications. In: Pocek, K.L., Arnold, J. (eds.) IEEE Symposium on FPGAs for Custom Computing Machines, pp. 302–303. IEEE Computer Society Press, Los Alamitos, CA (1998)
10. Shaw, A.: Communicating real-time state machines. IEEE Trans. Softw. Eng. **18**(9), 805–816 (1992). http://doi.ieeecomputersociety.org/10.1109/32.159840
11. Zuberek, W.M.: Performance evaluation using unbounded timed petri nets. In: PNPM'89, pp. 180–186 (1989)
12. Selic, B.: Using uml for modeling complex real-time systems. In: Proceedings of the ACM SIGPLAN Workshop on Languages, Compilers, and Tools for Embedded Systems, LCTES '98, pp. 250–260. Springer, London, UK (1998). http://dl.acm.org/citation.cfm?id=646905.710490
13. OMG: Unified modeling language (uml) resource page. http://www.uml.org/ (2015)

Chapter 6
Specification PEARL Security

As CPS are closely related to the physical processes they are part of, the validity and accuracy of the sensing process and the data collected during the process has to be ensured. Another important aspect of CPS is that they are networked by nature. This not only allows them to form networks for data fusion and delivery to back-end entities, but also to take coordinated response actions based on the data collected. Hence the data transfers among entities need to be secured. The third aspect of CPS is their data storage, where they must rely in part on the locally stored data and for the other to get reliable data from their networked and back-end entities. Hence, their data stores need to be kept secure and consistent.

The ultimate goal of Specification PEARL methodology is to provide CPS designer with capabilities for holistic design with safety and security features, thus making the final designs inherently safe and secure. In this chapter the focus is on the introduction of security mechanisms into the designs by means of known methods and mechanisms.

6.1 Design for Security

As already mentioned in the Introduction, the three relevant aspects of security that address the previously listed security concerns comprise:

- sensing and communication security,
- actuation control and feedback security, as well as
- storage security.

© Springer International Publishing Switzerland 2016
R. Gumzej, *Engineering Safe and Secure Cyber-Physical Systems*,
Studies in Computational Intelligence 632, DOI 10.1007/978-3-319-28905-2_6

6.1.1 Sensing and Communication Security

There are mainly two issues with CPS' sensing and communication security:

1. Sensing Security needs techniques to authenticate physical stimuli, so that any data measured in the physical processes can be trusted.
2. Communication Security needs the development of protocols to secure both inter- and intra-CPS communication from both active (interferers) and passive (eaves-droppers) adversaries.

In order to achieve sensing security all incoming signals need to be authenticated. For analog lines this is relatively easy—in case they have not been disconnected and reconnected there is no risk for false interpretation, since their origin is known. In case they have been reconnected the only way to secure the line is to inspect it and make sure it is inaccessible to anyone but trusted entities. Authentication of wireless connections is relatively complex since it requires bilateral authentication of both communicating entities. Hence, it requires a protocol, which ensures that the entities are identified during the establishment of a connection and that every connection is ended after the data transfer has taken place. The latter is to prevent stolen identities. During data transfer, the data needs to be encrypted in order to prevent eavesdropping.

6.1.2 Actuation Control and Feedback Security

Actuation Control and Feedback Security refers to ensuring protection of the control systems in a CPS which provide the necessary feedback for effecting actuation. We need to ensure that no actuation can take place without appropriate authorisation during the semi-active or active modes of operation. The authorisations have to be specified dynamically as the requirements for CPS change over time.

In order to ensure proper authorisation, the state guard needs to be extended with authorisation state. After every transaction the system ought to be returned to authorisation state to prevent stolen identities.

6.1.3 Storage Security

Once data have been collected and processed, they may be required to be stored over time for future access. Any tampering of these stored data can lead to errors during planning. Storage Security involves developing solutions for securing stored data in CPS platforms from physical or cyber-tampering.

The asymmetrical architecture can be considered inherently safe from the application point of view, since the control system's data structures are inaccessible to applications. On the other side, every stored data would need to be authenticated and every transmitted data should be identified in order to assure proper authenticity of data which should preferably be stored with the data. In a cloud data should be synchronised with their original data based on its origin tag. In case of data discrepancy, new data should automatically be transferred from the origin to all its mirror locations following the previously mentioned authentication procedure.

6.2 Securing Identification and Communication

Confidentiality, Integrity and Availability (CIA) are the three fundamental constituents of security. They apply to all CPS applications. To address this issue in a consistent manner, considering CPS life cycles and operation modes RFID technology will be taken as an example, since the various classes of RFID devices with their capabilities reflect the CPS operation modes. Since contemporary mobile devices, representing typical front-end CPS, are equipped with Wi-Fi, RFID and/or Near Field Communication (NFC) interfaces and represent RFID class 4 and 5 devices, the described solutions can be considered representative for secure identification and communication among front-end and back-end CPS.

6.2.1 RFID Security

RFID technology offers important advantages over conventional identification technologies, e.g. barcodes. However, they are also associated with various types of security drawbacks, originating from its vulnerability especially to eavesdropping and man-in-the-middle attacks. Hence, soon after its wider introduction, the need for lite security solutions has arisen in order to provide RFID applications with "Pretty Good Privacy (PGP)" protection. The reason for "lite" security solutions was the fact that most RFID devices, especially RFID transponders, offer very limited processing capabilities and hence cannot handle "strong" encryption algorithms in real time. On the other hand, RFID tags often contain sensitive information that should be readable only by dedicated readers, paired with the tags for the sake of security (e.g. remotely unlocking a car and disabling a car's engine immobilizer).

For most RFID applications the 3DES (Triple Data Encryption Algorithm) symmetrical cipher is considered sufficient. Here, RFID tags are paired with dedicated readers (e.g. ATMs to perform secure bank transactions). However, considering future RFID applications, including CPS with ubiquitous sensing, asymmetrical PGP-like secure identification and communication are likely to replace symmetrical ciphers. As the need for trusted readers shall arise, especially for ones that shall not only be solely authorised for reading certain groups of tags, but more importantly, for ones that shall be solely allowed to add information to their databases, they are considered mandatory.

Table 6.1 Proposed encryption mechanisms by RFID classes

RFID class	Transponder/Encryption mechanism	Reader keys
Class 1	Transponder ID is sent to the reader encrypted with the reader's public key; the reader uses its private key to decrypt the message	ID/Private key
Class 2 (Gen. 2)	Session key is generated and sent to the reader encrypted by its public key; the reader's message is signed by its public key encrypted by the session key	Session key/Private key (signed reader messages)
Class 3	Session key is generated, transponder's private key is used to sign outgoing messages; the reader's message is encrypted by the transponder's public key	Private key/Public key (signed transponder messages)
Class 4	Transponder's private key is used to sign outgoing messages; the reader's private key is used to sign its messages	Private key/Private key (signed transponder and reader messages)

Table 6.1 lists the appropriate encryption techniques to be used for different classes of RFID devices. Class 0 tags are intentionally left out due to their lack of any computing capabilities, rendering them unsuitable for any security-sensitive applications.

Class 1   Class 1 RFID tags are widely used for different purposes in the most common application areas. Usually, they implement some kind of encoding (e.g. Manchester) for the sake of data transfer safety. These tags are mainly meant for mass identification. Hence, no encryption is present. Since these tags are for read-only operation, the data on them do not need to be secured. One might wish to make sure, however, that only entitled readers can read their data in order to prevent identity thefts. This could easily be achieved by encrypting the transmitted data with the receiver's public ID. Only authorised receivers would then be able to receive correct data, while proprietary readers would only get ciphers.

Class 2   The more advanced class 2 RFID tags offer the possibility to add tracking or handling information to existing transponder data. In order to secure bidirectional transmission, session keys may be generated by a computationally light encryption procedure in order to secure the data transferred. The reader's public key should be used to authenticate and to authorise the reader when performing changes to the data stored in a transponder.

Class 3   Class 3 RFID devices are mainly used for admission control and authorisation as well as for process control. Since a transponder needs to be authorised to do something, which is then logged at the reader side, unique transponder identification is required. In order to secure bidirectional data flow and ensure proper authorisation, information sent needs to be accompanied by a digital signature. For information received the transponder's public key is sufficient, since it shall be able to decipher the message using its private key.

Class 4/5 The class 4 RFID devices are able to communicate among each other and with compatible devices. Hence, here is a need for strong encryption in both directions. Both ways data have to be properly signed (by private keys) in order to represent meaningful and authorised input. In connection with the concept of ubiquitous computing and the Internet of Things (IOT) the novel class 5 of active RFID transponders was introduced, which may act as readers as well. Since, from the CIA point of view, their operation behaviour does not differ significantly from that of class 4 devices, we consider them simply as an upgrade to this class of devices requiring mutual authentication for secure communication.

Meeting the "lite encryption" requirement according to diverse capabilities of RFID transponders (as indicated in Table 6.1), which are mainly due to limited processing power, number of logic gates and capability of rewriting or adding data to RFID transponders, in the sequel, two protocols are presented, one for *secure identification* and the other one for *secure communication*.

6.2.2 Secure Identification

As rather often only one-way authentication (e.g. for transponders of class 1)—sensing security—is needed, the *secure identification* protocol (Table 6.2), which provides a PGP-style authentication of a transponder, is introduced. With this identification protocol a transponder ID is verified by its digital signature and a session key. According to the protocol, a transponder generates a session key, encrypts it with the reader's public key and sends it to the reader. The reader receives the session key, which the transponder then uses to encrypt its own ID. Hence, only authorised readers can meaningfully decrypt the transponder's ID, using the provided session key. This protocol constitutes one-way authentication, where the transponder merely identifies itself to an authorised reader.

As a result of transponder authentication, the associated receiver action may be authorised. Hence, every CPS operation requiring automated authorisation should be protected by means of this protocol.

Table 6.2 Secure identification protocol

Transponder		Reader
Session key generation		
Sending the session key to the reader, encrypted with its public key	>	Decryption of the session key with own private key
Sending the transponder ID to the reader, encrypted with the session key	>	Decryption of the ID with the session key

6.2.3   Secure Communication

In cases, where mutual authorisation among transponder and reader is required for secure communication, the *secure communication* protocol as presented in Table 6.3, is introduced. The first three steps comprise identification and are identical to the *secure identification* protocol. In the sequel the reader identifies itself and addresses the transponder with the received ID to provide information. Since the usual relation of a reader towards a transponder is one-to-many, only the transponder, who was last identified, is addressed. In the further steps of the protocol secure exchange of transponder data and reader data takes place and transponder data is updated. Here, the protocol and session key validity ends. In order to start a new transmission the protocol needs to start from the beginning. When authenticating the reader, the transponder uses reader's public key to decipher the message that was encrypted with the reader's private key, but since public/private keys are symmetrical, this is possible and renders the correct result, provided listening transponder is the one, who initiated the communication with the session key. To ensure data consistency, the session key length needs to match the transmission block size enabling single-cycle encryption/decryption. The secure communication protocol is meant for secure bidirectional communication and should be used with transponders of class 3 and above.

Secure identification and communication protocols are appropriate wherever there is a need to provide CIA for communication between transponders and readers. The

Table 6.3 Secure communication protocol

Transponder		Reader
Session key generation		
Sending the session key to the reader, encrypted with its public key	>	Decryption of the session key with own private key
Sending the tag ID to the reader, encrypted with the session key	>	Decryption of the ID with the session key
Decryption of the tag ID with the session key and the reader's public key	<	Sending the tag ID encrypted with the reader's key, encrypted with the session key
Comparison of the received tag ID with own ID—if they match, continue, otherwise start from the beginning	–	
Sending tag ID and any additional information to the reader, encrypted with the session key	>	Decryption and processing of the received information with the session key
Add/modify tag information, decrypted with the session key	<	Sending tag ID and additional information to the tag, encrypted with the session key

The proposed encryption/decryption function is Boolean antivalence (XOR), since it is simple and symmetrical:

$$ccyphertext = XOR(text, session\,key) \leftrightarrow text = XOR(cyphertext, session\,key)$$

identification protocol ensures that only authorised readers can identify transponders, and provides resilience to eavesdropping, replay attacks, man-in-the-middle attacks and even against brute-force attacks, provided the session key length is sufficiently long. The secure communication protocol, on the other hand, provides the same level of security, not only in identifying both transponders and readers, but also in protecting the communication between them.

6.3 Securing Operation

For securing CPS operation authentication and authorisation procedures are extensively used in order to ensure that every operation performed by the system is properly authorised. There is a wide spectrum of applications that require this type of security, mainly related to distributed SCADA and PCS systems. Examples of such systems include, but are not limited to, remote traffic supervision, power plant supervision, waste water treatment supervision and also mobile health or telemedicine applications.

6.3.1 Biometric Security

Due to increasing demand on health care in developing countries, including high population growth, high burden of disease prevalence, lacking health care workforce, large numbers of rural inhabitants and limited financial resources to support health care infrastructure and enormously rising accessibility through cellular networks and the Internet, have motivated the development of mobile health and telemedicine applications. Mobile health or m-health is known as a provider of medical and public health services by means of mobile devices such as smart phones, tablets, mobile gear and wearable medical devices. In order to fully utilise wireless communication capabilities between wearable medical devices and physicians at back-end terminals and servers, the concept of Body Sensor Networks (BSN) was proposed in 2002 [1].

BSNs constitute a kind of wireless sensor network around human bodies and have great potential to be the main front-end platform of telemedical and mobile health systems. Therefore, their development is currently being strongly pushed forward to keep pace with the continuously rising demand for personalised health care. Comprised of sensors attached to a human body for collecting and transmitting vital signs, a BSN facilitates the joint processing of medical data collected at different parts of the body at different times for purposes of resource optimisation and systematic health monitoring. In a typical BSN, each sensor node collects various physiological signals in order to monitor the patient's health status regardless of the patient's location and transmits all gathered information in real time to a medical server or to physicians in charge [1]. Following this procedure complex telediagnostics and limited medical treatments are possible with appropriately trained medical personnel

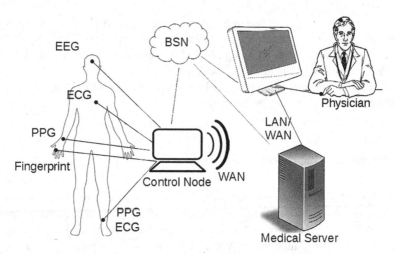

Fig. 6.1 An application scenario of body sensor networks

or the patient only. In case an emergency is detected, the physician shall immediately inform the patient through the computer system by providing appropriate messages and alarm the closest medical team to come to aid. Hence, BSNs are preferred in monitoring and treating patients in environments lacking medical doctors, such as homes and workplaces. Figure 6.1 presents a simplified example of a BSN application scenario in a mobile health system. Sensor nodes on or inside the human body and a Control Node are interconnected to form the BSN. Medical information, collected by different sensors of the BSN, are sent to the Control Node for data fusion and preparation to be forwarded to a central Medical Server for further analysis, or directly via various forms of wireless communications, such as Wireless Personal Area Networks (WPAN), Wireless Local Area Networks (WLAN) or Wide Area Networks (WAN), to physicians for treatment.

In medical applications, wireless networks have to provide high levels of reliability in order to guarantee security of patient's information and privacy of health care history. To ensure security of overall mobile health systems, as an important part, BSNs should be protected from attacks such as eavesdropping, injection and modification. This is a non-trivial task, however, due to rather limited processing, memory and energy resources as well as due to the lack of appropriate user interfaces for unskilled users, the longevity of devices and global roaming for most sensor nodes [1].

Symmetric cryptography, for which communicating parties must exchange shared secret keys via invulnerable key distribution facilities prior to any encryption process, is a promising approach to relieve the strong resource constraints holding for BSNs. Existing key distribution solutions for large-scale sensor networks, such as random key pre-distribution protocols and polynomial pool-based key distribution [2], require some form of pre-deployment. Given the progressively increasing deployment of

BSNs, however, these approaches may potentially involve considerable latency during network initialisation or any subsequent adjustments due to their need for pre-deployment. In addition, it obviously discourages users to newly configure initial keys any time when there is a need to add or change a body sensor in order to ensure that new sensors can securely communicate with the existing ones. Therefore, new key distribution solutions are desirable for BSNs which do not require any form of initial deployment to provide plug and play security.

It is well known that physiologically and biologically the human body possesses its own transmission capabilities such as the blood circulation system. Thus, it is a good idea [3] to make use of these already available communication pathways to secure BSNs for telemedical or m-health applications, as for collecting medical data nodes of such BSNs could comprise biosensors with physiological characteristics uniquely representing an individual. If such intrinsic characteristics could be used to verify whether two sensors belong to the same individual, the use of physiological signals to identify individuals and to securely transmit their private encryption keys would become feasible and save resources. Building upon this general idea, a family of low-expense and resource-efficient security solutions based on time-variant physiological signals has been proposed for BSNs with the dual purpose individual identification and secure key transmission. This approach differs from traditional biometrics, where the physiological or behavioural characteristics are static and merely used to automatically identify or authenticate individuals [1]. The traits utilised in traditional biometric systems are expected to have characteristics such as universality, distinctiveness, permanence, effectiveness or invulnerability, while the physiological characteristics should be dynamic at different times to ensure the security of key transmissions in BSNs. In the biometrics-based security solution depicted in Fig. 6.2, physiological signals of the human body such as electrocardiogram (ECG) and photoplethysmogram (PPG) are used to generate an Entity Identifier (EI) of each node for its identification, and for protecting the transmission of medical data (MD) by a data encryption/decryption processes. Its verification is based on the fact that EIs generated simultaneously from the same subject share great similarity, while those generated non-simultaneously or from different subjects exhibit significant differences.

Biometrics can establish personal identities from the moment on patients enter the care of a physician or a medical facility. Subsequently, these identities can be transmitted accurately and securely throughout health care information systems. Biometrics can be used to ensure that only authorised (medical) personnel is permitted to access a patient's medical records and sensitive hospital facilities, such as nurseries and operating rooms, and that prescribed medications are delivered to the proper patients. The technology will foster positive health care identification and will enhance the secure use, storage and exchange of personal health records, and secure medical treatment by telemedicine applications. It will provide for proper authorisation to any health sensitive operation through positive identification of responsible medical doctors and other medical personnel.

Fig. 6.2 Workflow of a
biometrics-based security
solution

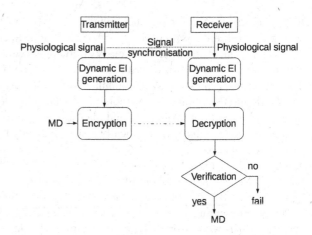

6.3.2 One-Time Pad

One-time keys are often used to encrypt messages performing atomic transactions over the Internet, e.g. in Internet banking. Provided the keys are sufficiently long, namely as long as the messages encrypted, this so-called "Vernam-cipher" is information-theoretically secure, i.e. unbreakable with systematic methods or by brute-force attacks. This is achievable, if the probability of an arbitrary ciphertext for a given plaintext is equal—this makes it impossible to decipher the plaintext based on the intercepted ciphertext. According to the information-theoretical theorem by Shannon [4], a cryptographic system is perfectly safe only if the number of possible keys is at least as high as the number of possible messages. This is achievable by equal lengths of the encrypted messages and their encryption keys. Current cryptographic techniques are usually based on the keys that are used to encrypt multiple messages during a longer period of time, which makes them cryptographically vulnerable. Only cryptographic algorithms for which currently available computing power is not sufficient to check all possible keys are considered cryptographically safe. Hence, only one-time keys are sensible in the perspective and were used as basis for our authentication and authorisation protocol for CPSs.

To achieve sensing and communication security for mobile CPSs, their nodes need to be equipped with relatively long bit strings (keys) the size of a message. Upon connecting a node to an authorisation server, a one-time key is delivered to the node via a secure channel. Any of such keys may be used only once, and needs to be replaced by a fresh one with every new communication.

The authentication and authorisation protocol for CPSs reads as follows:

- Before a node commences a secure data transmission with another node (terminal or server), it registers itself with the authorisation server where it gains its identifier (ID) and authorisation code (AC).
- Upon commencing a secure transmission the node transmits its data transfer request, containing its identifier and request type, to the other node.

- Upon receipt of the data transfer request a node first checks the plausibility of the incoming request by consulting the authorisation server. If the plausibility data are inconsistent, the communication is terminated.
- If the sending node was positively identified (authenticated), the receiving node sends an encrypted authorisation request to all its active and accredited nodes, encrypted with the current one-time key for the communication with the requesting node (AC). Herewith, this node is solely capable to successfully decrypt the authorisation request.
- The sending node replies to the authorisation request with its entity identifier (EI), e.g. a fingerprint of the operator, patient etc., which the receiving node compares with the contents of its database before authorising the node to access its data and services. If the comparison fails, the requesting node is considered manipulated and communication with it is terminated.

After successful authorisation the communication among nodes may commence—a node can transfer data to another node, access its data and services using the current one-time key for their communication until the communication protocol ends. In case the message transfer is interrupted, a new connection must be set up using the same security protocol. Communication needs to be closed automatically after a certain time period, allotted for one session, expires to prevent hijacking.

The combination of one-time (AC) and biometric keys (EI) for mutual authorisation of any single communication is the most tamper-proof method of secure authentication, authorisation and communication within CPSs with their various entities, transponders and servers. Hence, it is the preferred method of secure communication for CPSs.

The CPS nodes need to be tamper-proof in order to prevent their detachment from the environment they control and misuse of the information they contain. In case this happens they need to implement the functionality of a tamper switch, which would detect this event and prevent its further authentication and authorisation by erasing its authentication (ID) and authorisation keys (AC). Depending on the environment and nature of their application these tamper switches may be physical (triggered by detachment of a control circuit from a device) or software (triggered by attachment of another circuit triggering unprotocolled actions, e.g. memory sweeping).

6.4 Securing Storage

In order to secure CPS storage every peace of data in shared storage needs to be tagged with a timestamp and authentication data. In order to synchronise data among associated CPSs the timestamp tells if the data within the node needs to be updated to the current status. The node then needs to authorise another node to upload its data to this node following the previously described authentication and authorisation protocol. Upon update the identity of the data origin node is stored with the data for later updates. The data need to be indexed by identity in order to be easily accessible during the update process.

6.5   Security Shell

It is crucial for the before mentioned security mechanisms to be systematically integrated into the CPS life cycle. A means for systematic integration of safety mechanisms in the form of a safety shell has already been presented. Since one cannot speak about security without safety, the security shell shall build on the safety shell.

CPS life cycles can usually be broken down into three phases:

1. *integration*, being associated with its identification and authentication with its surrounding networked nodes
2. *operation*, being associated with the functioning of a CPS node in its surroundings, according to its specifications (imported and exported functions) and
3. *disintegration*, being associated with the CPS nodes being down for maintenance or leaving the perimeter to join another surroundings.

As already mentioned, their workflows can be broken down into four main functions:

1. *Monitoring*, deals with gathering data from the environment, protected through I/O monitor for safety and added security; depending on the type of data and device, these data may also be (temporarily) stored in the CPS in its secure shared storage; the safety and security shells function in conjunction to provide for CIA of the stored data.
2. *Analysis* deals with analysing the data, collected during monitoring, to determine whether the physical process is meeting the specified criteria; the process relies on the stored data, being secured by the previously mentioned I/O monitor.
3. *Planning* is important in situations when the criteria are not satisfied; here, corrective actions are determined and carried out under surveillance of the state monitor with exception handling and timing monitor to assure CPS node's correctness, timeliness and availability.
4. *Execution* deals with the actuation of actions determined during the planning phase; it can take many forms from changing the cyber-behaviour of the CPS to controlling the physical process itself; in conjunction with the already mentioned state and timing monitors the integrity and availability of this process can be assured, confidentiality however needs to be assured by proper authorisation of the system's users through the described authentication and authorisation protocol.

According to CPS node's specifications the safety shell's state monitor states need to be organised accordingly to reflect its workflow:

0 initial state; in this state the CPS node is initialised and authenticated with its surroundings stating its needs and capabilities.
1 monitoring state; in this state the CPS node is gathering data from its environment—trusted data (transferred through secure channels) are stored in its data store for further processing.
2 analysis state; in this state the CPS node is analysing the data, producing results and storing them in its own or shared storage; since these data are to be used

in the next state, they need to be checked for coherency and consistency where both—the data authenticity as well as input and output data ranges—need to be checked to provide safe and secure results.

3 planning state; in this state the data from the previous state are used to decide on the necessary actions, performed by the node itself, its surrounding nodes or its back-end system; at this stage the correctness of the decision-making tasks is crucial, hence from security point of view they should be protected from any form of malware by assigning them memory areas, being inaccessible to the outside of the CPS and to perform their updates from properly authenticated sources only through secure communication channels.

4 execution state; in this state the planned actions are executed but beforehand appropriate authorisations are sought from the node's user and/or the back-end system.

Unless the node's life cycle forces it to a halt position (e.g. for maintenance) the node's initial state is used only once. The rest of the states are chosen consecutively and cyclically in cycles comprising 2 or 4 states depending on the CPS node's operational modes.

As already mentioned, a CPS can operate in one of the three possible modes:

1. *Passive*: in this mode CPS act as information gathering platforms only, and solely monitor their environment, gather data and prepare them for processing.
2. *Semi-Active*: in this mode CPS monitor their environments (physical aspect) and analyse the data; if they detect some criteria not to be fulfilled, they execute indirect actions to change their own behaviour (cyber-aspect), so that the criteria can be satisfied.
3. *Active*: in this mode CPS monitor their physical environments and analyse the data; if they detect some criteria not to be fulfilled, they execute direct actions to modify the behaviour of the physical environments, so that the criteria are satisfied.

Depending on the CPS node's specifications the before mentioned states are worked off in the following cycles:

1. *Passive*: states 0 and 1.
2. *Semi-Active* and *Active*: states 0, 1, 2, 3 and 4; according to the operational mode the difference occurs in state 4, where the node changes its own behaviour or changes the behaviour of the environment, however this does not change the cycle.

Hence, by systematic integration of the previously mentioned security policies with the safety shells mechanisms a safe and secure CPS execution environment can be obtained.

6.6   Security Level Specification

To assure any security level according to the previously described standards on CPS security, one cannot solely protect the individual segments of CPS (e.g. I/O). Hence, based on the level of security one wishes to achieve, appropriate security measures need to be integrated into the CPS designs. Table 6.4 summarises the mechanisms according to CPS operational modes and security levels. To comply with SL1, positive identification of incoming requests or data is necessary. This security level is unsuitable for semi-active or active operational modes. Hence, the security mechanisms in this case are non-applicable. To comply with SL2 unique authentication of users or programmes accessing our CPS in required. This level is already applicable for semi-active operational modes, since it requires only internal authorisation to perform changes within the CPS. To comply with SL3 multi-factor authentication and authorisation is required. This level of security is applicable also to the active mode of CPS operation where it relies on mutual authentication and authorisation among trusted entities on a trusted network. With SL4 the highest requirements on authentication and authorisation of user or programme access to the CPS via an untrusted network are required. Here for every operation multi-factor authentication and authorisation is required.

To provide for safety and security appropriate safety and security mechanisms should be combined in order to provide for the required level of dependability. Table 6.5 summarises the combined affect of safety and security levels on CPS dependability. From the above consideration, we can deduce that basic CPS applications on SL1 and SL1 only cannot be attributed any dependability. Dependable passive mode CPS are considered applications with at least SIL3 and SL2 compliance. Dependable active mode CPS are considered applications with at least SIL3 and SL3 compliance. For safety and security critical applications SIL4 and SL4 compliance is necessary. It can be achieved by fully implementing the safety shell with all applicable security mechanisms for the application.

To specify the pertaining safety and security properties of a CPS project the respective SIL and SL levels are to be noted. In Specification PEARL they specify the rigour in which the mechanisms, provided by the methodology, are used and checked.

Table 6.4 Security mechanisms by CPS operational mode and security level

	Passive	Semi-active	Active
SL1	Identification	Non-applicable	Non-applicable
SL2	Unique authentication	Unique authentication and authorisation	Non-applicable'
SL3	Multi-factor authentication	Multi-factor authentication and authorisation	Mutual authentication and authorisation
SL4	Multi-factor authentication	Multi-factor authentication and authorisation	Mutual multi-factor authentication and authorisation

Table 6.5 The combined effect of the chosen SIL and SL on CPS dependability

	SL1	SL2	SL3	SL4
SIL1	Passive mode CPS without safety or security attributes	Non-applicable	Non-applicable	Non-applicable
SIL2	Fail-safe passive mode CPS	Fail-safe and secure passive mode CPS	Non-applicable	Non-applicable
SIL3	Non-applicable	Safe passive mode CPS	Safe and secure semi-active mode CPS	Safe and secure active mode CPS
SIL4	Non-applicable	Safety critical passive mode CPS	Safety and security critical semi-active mode CPS	Safety and security critical active mode CPS

References

1. Miao, F., Bao, S.D., Li, Y.: New trends and developments in biometrics: physiological signal based biometrics for securing body sensor network. In Tech (2012)
2. Liu, D., Ning, P.: Establishing pairwise keys in distributed sensor networks. In: The 10th ACM Conference on Computer and Communication (2003)
3. Poon, C.C., Zhang, Y., Bao, S.D.: A novel biometrics method to secure wireless body area sensor networks for telemedicine and m-health. IEE Commun. Mag. 73–81 (2006)
4. Shannon, C.: Communication theory of secrecy systems. Bell Syst. Tech. J. **28**, 656–715 (1949)

Chapter 7
Evaluation of the Methodology

7.1 Design for Correctness and Timeliness

The Specification PEARL methodology enables early reasoning on integrating systems. At the same time, the hierarchical structure of system architectures enables top-down stepwise refinement in design. Specification PEARL models are finally deployed to application prototypes for execution on specified target architectures. They are fine-tuned and verified for their temporal properties by co-simulation. Architecture descriptions are comprehensive and syntactically clean enough to be used as inputs to configuration managers. They are also user-readable and can, therefore, be used for audits as part of the program documentation. The formal description of the program tasks in form of timed state transition diagrams was chosen with respect to the method foreseen to verify and validate designs as well as regarding the safety issue.

Designers first set up the logical structure of a system and its components/parts to be described in detail at different levels. When at least the logical hardware architecture is set up, software units (collections) may be associated with it. The design can also be started from the software point of view, and the mapping can be carried out after the stations of the hardware architecture have become available. Software ports can be mapped to hardware interconnection lines as soon as they are configured.

Possible incompatibility of parameters is checked immediately or while creating the architecture description in the language Specification PEARL. Before commencing the verification and validation of the system modelled, it is checked on completeness and parameter compatibility, since a consistent, unambiguous description is needed for validation. Consistency checking ensures that the minimum preconditions for a successful verification and validation are fulfilled.

The validation of a design proposed can show its weaknesses in the hardware or the software parts. The system architecture description's detailedness directly influences the quality of co-simulation results. At the end, auditors evaluate the

© Springer International Publishing Switzerland 2016
R. Gumzej, *Engineering Safe and Secure Cyber-Physical Systems*,
Studies in Computational Intelligence 632, DOI 10.1007/978-3-319-28905-2_7

results of co-simulations together with the designers. Considering their observations, the system design at hand is accepted or declined, and it is decided whether and which modifications are necessary.

7.2 Design for Safety

The Specification PEARL methodology with its concepts, constructs and mechanisms fulfils the regulations holding for the safety integrity level SIL 3 and, for certain safety-related issues, also for SIL 4 according to the standard IEC 61508. The individual guidelines listed in Table 1.1 are addressed as follows:

- Use of coding standard—the Specification PEARL methodology builds on standard PEARL [1, 2] and PEARL for distributed systems [3]. In order to ease the deployment on microcontroller-based architectures also cross-compilation to ANSI C and appropriate HaRTOS and CM libraries written in C were devised.
- No dynamic variables—automatic code generation enables strong typing as well as prevention of dynamic variables, since all relevant project limitations on hardware as well as on software can easily be changed through co-design.
- Structure-based testing—consistency and coherency checking are provided by the underlying tools, enabling model-based consistency checking as well as Specifications-based coherency checking during co-simulation.
- Failure modes, effects and criticality analysis—the safety shell and CM mechanisms enable to compile all conceivable states or modes a system can assume, and provide error-handling mechanisms for failure states, which allow for graceful degradation and return to normal operation mechanisms.
- Formal methods modelling—the textual modelling language is based on strict syntax rules. The use of timed state transition diagrams enables formal descriptions of program tasks.
- Response timings and memory constraints—hardware/software co-design enables to estimate the memory consumption. Since dynamic variables are disabled, memory consumption is unlikely to exceed preset values. Timeliness analysis by co-simulation provides the opportunity to determine sensible response times in all conceivable situations.
- Performance requirements—hardware/software co-design enables feasibility validation through timeliness analysis by co-simulation that can be proven by temporal analysis on the target system.
- Finite state machines—timed state transition diagrams are finite state machines with time limitations. The addition of time limitations did not change their basic behaviour, whilst the time-out mechanism makes no provisions for the rigour of their handling except for the fact that they are handled.
- Boundary value analysis—co-simulation uses boundary value analysis to cover as much time space as possible. To ensure temporal feasibility of designs produced

minimum as well as maximum execution times are considered for each state a task can assume.

- Walk-throughs/design reviews—are carried out to check the foreseen operational states of the safety shell according to the established risks of failures.
- Software module size limit—is established by setting the properties of appropriate storages of stations to which collections are mapped. In addition, the collections are designed with fixed numbers of modules and tasks, so their memory requirements are known at design time.
- Information hiding/encapsulation—the co-design approach supported by the Specification PEARL methodology ensures information transparency among hardware and software designs at exactly defined spots. The modular approach to encapsulate tasks also supports information hiding by defining local variables and procedures/tasks.
- Fully defined interface—by following the Specification PEARL methodology systems can be engineered from any aspect of design, use and deployment. The methodology covers the entire life cycle including re-engineering.

The Specification PEARL methodology corresponds to the safety life cycle (see Fig. 1.1) and its planning, design, verification and validation as well as implementation methods according to the IEC 61508 guidelines for SIL 3 or SIL 4. At least one of the recommended methods (see Table 1.1) is used in each phase of the system life cycle:

1. Concept: all information on the EUC (physical, legislative etc.) relevant to the following steps needs to be compiled. This step is obviously well covered by the methodology, since the system is designed holistically—from the architectural (hardware and software) and functional perspectives. The legislative demands are not covered by the methodology, since they represent parts of the specification being non-functional and to be considered separately by audits.
2. Overall scope definition: specification of the *hazards* and *risks* (e.g. process hazards, environmental hazards) for safety-related devices need to be compiled in the specification to enable later validation.
3. Hazard and risk analysis: is performed to determine the *hazards* and hazardous events or sequences of events relevant for the EUC (and the EUC control system) for all foreseeable cases including fault conditions and misuse, and to determine the associated *risks*. They are considered when building the *safety shell* of the system.
4. Overall safety requirements: are compiled to develop a specification for the overall safety requirements in terms of the safety functions requirements and *safety integrity* requirements in order to achieve the *functional safety* required. Since a system's functional states heavily depend on this issue, its design needs to consider all possible states it may assume, and define transition conditions among them that enable, e.g. graceful degradation in failure conditions and return to normal operation afterwards.

5. Safety requirements allocation: imposes the safety functions from the specification produced previously on the designated E/E/PE or on other technology safety-related systems as well as external risk reduction facilities, and allocates a *safety integrity level* to each *safety function*. This is done in the safety shell by breaking down risks and handling within temporal-, I/O- and state-related monitoring.

6. Overall operation and maintenance planning: is carried out for E/E/PE safety-related systems to ensure that functional safety required is maintained during operation and maintenance. As for *operation*, it is done in form of the safety shell. The *maintenance* part, however, is out of the Specification PEARL methodology's scope and must be considered separately, unless it concerns the automatic updating feature, being part of the system specification.

7. Overall safety validation planning: facilitates the overall *safety validation* of E/E/PE safety-related systems. As described previously in the context of the Specification PEARL methodology, it is provided in form of planning consistency and coherency verification as well as dependability and predictability validation.

8. Overall installation and commissioning planning: is carried out to ensure that the required functional safety of E/E/PE safety-related systems is achieved during these phases. It is also out of the scope of Specification PEARL methodology, since the produced systems are considered prototypes.

9. Safety-related systems (E/E/PES) realisation: comprises the creation phase of E/E/PE safety-related systems where the specified functional safety and safety integrity requirements have to be obeyed. This is realised by the safety shell implementation.

10. Safety-related systems (other technology) realisation: comprises the creation phase of other safety-related systems where the specified functional safety and safety integrity requirements for these specific technologies have to be obeyed (outside the scope of this standard). This is realised by the safety shell's I/O guard, since it concerns the I/O from/to these "foreign" devices.

11. External risk reduction facilities' realisation: comprises the creation of external risk reduction facilities to meet the requirements specified for the safety functions and their safety integrity (outside the scope of this standard). They are planned in the previous step. Since they are not part of the systems designed, however, they must be realised separately.

12. Overall installation and commissioning: comprises the *installation/ commissioning* phase of E/E/PE safety-related systems. This step represents the end of the Specification PEARL project life cycle.

13. Overall safety validation: is to validate that the functional safety and safety integrity requirements specified for E/E/PE safety-related systems are met. The safety shell may be tested by inducing predefined failure conditions.

14. Overall operation, maintenance and repair: functionally comprises intact operation, maintenance and repair of E/E/PE safety-related systems with respect to safety. When applying modifications to a system (prototype) its safety shell

needs to be updated as well, since new failure conditions may arise—this requires going back to step 3.

15. Overall modification and retrofit: is meant to ensure that the functional safety of E/E/PE safety-related systems is appropriate during and after these phases. When all functional and safety updates have been applied and safety validation has been performed (step 13), a system shall be put into function. Any subsequent modifications need to be done the same way—this requires going back to step 1.

16. Decommissioning or disposal: is meant to ensure that the functional safety of E/E/PE safety-related systems is appropriate during and after these phases both for EUCs and their control systems. The Specification PEARL methodology does not foresee an end of the project life cycle. Rather, it considers the previous one as the final step.

Since UML, being a prominent design methodology also for embedded (real-time) systems, is still lacking some of the features of Specification PEARL, a matching UML profile was defined. Also the semantic translation of timed state transition diagrams, used to model tasks in Specification PEARL, into UML state charts, was presented in the previous chapter. Hence, for UML-based design of CPS oriented at Specification PEARL the two methodologies can be combined in the framework of a stereotyped UML model, provided the corresponding configuration management classes and their associated architecture data structures are included in the application models. A UML safety shell pattern was devised to ensure safety and security in UML Specification PEARL projects.

7.3 Design for Security

Considering the autonomous nature of cyber-physical systems (CPS) there are four basic properties that need to be fulfilled in order to ensure their correct, timely as well as safe and secure operation:

1. Self-management: provides the self-controlling and self-updating functionality to CPS. It also preserves a stable, safe and secure operation mode of CPS.
2. Self-configuration: provides CPS with adaptability to environmental changes by smart reconfiguration options.
3. Self-healing: provides CPS with safety properties that enable them to diagnose, isolate and fix sources of their instability.
4. Self-protection: provides CPS with security properties by means of self-healing and intrusion protection by early risk detection and introducing appropriate protection mechanisms.

Properties 1 and 3 are addressed by the Specification PEARL's safety shell. The timing guard, state guard and exception handler ensure the system's stable and safe operation. Property 2 is addressed by the Specification PEARL's configuration management (CM) scheme. It provides for the system's adaptability to its and/or its

Table 7.1 Proposed security measures for CPS based on ISA-99.03.03, Draft 4

System requirement	SL
SR 1.1 The control system shall provide the capability to identify and authenticate all users (humans, software processes and devices). This capability shall enforce such identification and authentication on all interfaces which provide access to the control system (I/O guard) to support segregation of duties and least privilege in accordance with applicable security policies and procedures	1
SR 1.1 RE 1 The control system shall provide the capability to *uniquely* identify and authenticate all users (humans, software processes and devices). This capability shall enforce undeniability of authentication to enable authorisation of control system's actions	2
SR 1.1 RE 2 The control system shall provide the capability to employ multifactor authentication for human user access to the control system *via an untrusted network* (see 4.12, SR 1.10—Access via untrusted networks). For requests from the Internet multiple authentication criteria need to be fulfilled to provide plausible undeniability for the same purpose as with SR 1.1 RE 2	3
SR 1.1 RE 3 The control system shall provide the capability to employ multifactor authentication for *all* human user access to the control system. This capability is useful for providing peer-to-peer secure connections among CPS devices on the Internet	4

environment's state changes. Property 4 is partly ensured by the safety shell's I/O protection which filters out only the meaningful input data, so it can be processed in concordance with the system's specifications. In addition, here the described security mechanisms can be applied, especially with respect to authentication, authorisation and secure communication.

Considering the measures that have to be in place on system and component levels to assure the corresponding security levels (e.g. Table 7.1), the described security mechanisms need to be employed with increasing rigour and extent, resulting in different degrees of CPS compliance with the SL. The requirements SR 1.1 and SR 1.1 RE 1 in Table 7.1 are implemented by the "Secure Identification" (one-way authentication, cf. Table 6.2) protocol. The requirement SR 1.1 RE 2 in Table 7.1 is implemented by "Secure Communication" (mutual authentication, cf. Table 6.3). The requirement SR 1.1 RE 3 in Table 7.1 is addressed by the "Authentication and authorisation protocol for CPS" (cf. Sect. 6.3.2).

The validation of compliance of a Specification PEARL project with SIL and SL levels may be done automatically (within the associated CASE environment), semi-automatically where parts of the design are audited manually or completely manually where all parts of the design are audited by human inspection. The Specification PEARL CASE environment enables semi-automatic validation, where the mechanisms for providing safety and security are enabled or disabled in the design, according to the corresponding SIL and SL level settings of the project; however, the final inspection should be done by auditing.

7.4 Design for Licenseability

Considering the existing standards on quality of information systems (ISO/IEC 13236 [4] and related standards, e.g. [5, 6]), it may be said that they have been considered in the Specification PEARL methodology to provide CPS projects with a life cycle enabling the highest—managed—level of quality. General properties relating to system quality such as functionality, reliability, usability, efficiency, ease of maintenance and interoperability have been built into the associated design tool [7]. Here, the users', managers' and developers' views on a system are being accounted for in the process of system design, since they are integrated with system evaluation as additional criteria for selecting alternative designs.

As already stated in the section on design for safety, the Specification PEARL CPS designs can adhere to various levels of the four safety integrity levels (SIL1–SIL4), however the design methodology and CASE environment with their basic design constructs and libraries fulfil SIL4, i.e. the most critical one. Here all prescribed activities at different levels and phases of system development (e.g. coding standards, dynamic analysis and testing, black-box testing, failure analysis, modelling, performance testing, formal methods, static analysis and modular approach), which are desired or mandatory, and approaches, which are allowed or required in order to fulfil the requirements of SIL4, were accounted for. Hence, it may be stated that Specification PEARL designs comply with the rules from the standard IEC 61508 [8] for life cycle management of instrumented protection systems.

In the above section on design for security it is already stated which security measures were employed by the Specification PEARL methodology to achieve cyber-security for CPS at different levels. They have been selected based on the ISA 99 series of standards, which address the subject of cyber-security for industrial automation and control systems, namely ANSI/ISA 99.01.01-2007 [9], ANSI/ISA 99.02.01-2009 [10] and ANSI/ISA 99.03.03-2013 [11]. Hence, by the choice of concepts and models related to cyber-security, they also comply with the derived standards for cyber-security management systems issued by the IEC TC 65 WG 10, namely IEC 62443-1-1 [12] and IEC 62443-2-1 [13].

Based from the above statements, a conclusion is possible, that the methodology and supporting CASE environment itself and the resulting CPS designs are licenseable for quality, safety and security.

References

1. 66253, part 1: Basic pearl. Tech. rep., DIN (1981)
2. 66253, part 2: Full pearl. Tech. rep., DIN (1982)
3. 66253, part 3: Pearl for distributed systems. Tech. rep., DIN (1989)
4. Institution, B.S., for Standardization, I.O.: Implementation of ISO/IEC 13236: information technology: quality of service: framework. British Standards Institution. http://books.google. si/books?id=mpkgHAAACAAJ (1996)

5. ISO: International standard ISO/IEC 9126: information technology—software product evaluation—quality characteristics and guidelines for their use. Tech. rep., International Standard Organization (1991)

6. ISO: International standard ISO/IEC 9127: information processing systems–user documentation and cover information for consumer software packages. Tech. rep., International Standard Organization (1998)

7. Gumzej, R.: Holistic embedded control systems design with specification pearl. 1 CD-ROM. http://www.rts.uni-mb.si/misc/projekti/SPEARL/ (2006)

8. 65A, I.S.: Functional safety of electrical/electronic/programmable electronic safety-related systems. Tech. Rep. IEC 61508, The International Electrotechnical Commission, 3, rue de Varembé, Case postale 131, CH-1211 Genève 20, Switzerland (1998)

9. ANSI/ISA 99.01.01-2007, security for industrial automation and control systems part 1: terminology, concepts, and models. http://webstore.ansi.org/RecordDetail.aspx?sku=ANSI (2007)

10. ANSI/ISA 99.02.01-2009, security for industrial automation and control systems: establishing an industrial automation and control systems security program. http://webstore.ansi.org/RecordDetail.aspx?sku=ANSI (2009)

11. ANSI/ISA 99.03.03-2013, security for industrial automation and control systems part 3-3: system security requirements and security levels. http://webstore.ansi.org/RecordDetail.aspx?sku=ANSI (2013)

12. IEC TS 62443-1-1:2009, industrial communication networks—network and system security—part 1-1: terminology, concepts and models. https://webstore.iec.ch/publication/7029 (2009)

13. IEC 62443-2-1:2010, industrial communication networks—network and system security—part 2-1: establishing an industrial automation and control system security program. https://webstore.iec.ch/publication/7030 (2010)

Chapter 8
Conclusion

Different approaches to co-design, verification and validation of cyber-physical systems (CPS) have been discussed. New technologies for the development of CPS have become a must due to their increasing complexity and the expanding area of their use. In our research, hardware–software co-design, automatic programme code generation as well as verification and validation of CPS designs have been proposed in the form of the Specification PEARL methodology. Here, automatic verification, validation and retrofit are enabled by its CASE environment's co-simulation feature.

By utilising the holistic approach, finally, the designs can meet all specified requirements as a whole (in their hardware as well as in software parts), considering the desired functional correctness, timeliness, safety, security and licenseability. At the same time, the Specification PEARL co-design methodology offers the possibility of automatic documentation creation. A system as a whole may be observed from the start of the project from the user's, designer's and implementer's points of view at different levels of detail.

In this book the Specification PEARL methodology is proposed to co-design CPS, since it enables holistic co-design, verification and validation of CPS in order to enable their manageable and standards compliant design, development and deployment. In the long run, this approach has turned out to be also the least pricey and most sustainable one.

© Springer International Publishing Switzerland 2016
R. Gumzej, *Engineering Safe and Secure Cyber-Physical Systems*,
Studies in Computational Intelligence 632, DOI 10.1007/978-3-319-28905-2_8

Appendix A
Textual Architecture Description

A textual architecture description is composed of four divisions, each one describing another aspect of the architecture specification. Its graphical counterpart has two parallel layers, representing the HW and SW architectures, respectively. Here the complete set of the Specification PEARL language constructs is presented. The mentioned basic set is simply a limited version of the full set.

    ARCHITECTURE_description ::=

        'ARCHITECTURE' ':'
        ARCHITECTURE_descr
        ARCHITECTURE_divisions
        'ARCHEND' ';'

    ARCHITECTURE_descr ::=

        'NAME' ':' ARCHITECTURE_id
        [ARCHITECTURE_SIL_spec]
        [ARCHITECTURE_SL_spec] ';'

    ARCHITECTURE_SIL_spec ::=

        'SIL' ['1' | '2' | '3' | '4']

    ARCHITECTURE_SL_spec ::=

        'SL' ['1' | '2' | '3' | '4']

    ARCHITECTURE_divisions ::=

        STATION_division
        NET_division
        SYSTEM_division
        CONFIGURATION_division

© Springer International Publishing Switzerland 2016   91
R. Gumzej, *Engineering Safe and Secure Cyber-Physical Systems,*
Studies in Computational Intelligence 632, DOI 10.1007/978-3-319-28905-2

A.1   Station Division

In its station division a system's processing nodes are introduced, stating their most important characteristics. Each station in a system is uniquely identified. Stations are treated as "black boxes" with connections through their "INTERFACES". The "stations" are being interconnected by "drawing" "lines" between them.

To each station its state information (a "register" being monitored by the CM) is assigned. Depending on the current state of the station, the appropriate SW collection is chosen for execution by the station's CM.

Several types of stations have been defined, depending on the role they play in the overall system architecture. The default type is BASIC station, representing a general-purpose processing node. To be able to describe distributed and hierarchical (micro-kernel) architectures, additional types of processing nodes have been defined, viz. KERNEL STATION to represent RTOS stations, TASK STATION to represent pure application stations, COMPOSITE STATION to represent hierarchical architectures, and stations having no RTOS, but a CM knowing the rooting schema.

All stations must load the CM and may also load an RTOS in case an RTOS functionality is required. Stations are considered to be RTOS-less unless stated otherwise. Each station initially loads the CM, which also performs initial task loading and initialisation of the RTOS if present (initialising the Task Control Block (TCB) tables and scheduling initial tasks for execution).

The basic set of Specification PEARL constructs is mainly meant for small embedded systems, whereas the full set also addresses implementations of larger distributed and hierarchical architectures. In the full set, stations may be composed of substations to represent layered architectures. This feature is introduced in textual specifications by the "PART OF" construct.

STATION_division ::=

>   'STATION' ':'
>   STATION_descr { STATION_descr }
>   'STAEND' ';'

STATION_descr ::=

>   STATION_name
>   { STATION_attribute }
>   { STATION_state_id }
>   [STATION_PARTOF_opt]
>   [STATION_type_descr_opt]

STATION_name ::=

>   'NAME' ':' STATION_id ';'
>   STATION_id ::= [['(' Int_const_denot ')'] id | Int_const_denot]
>   STATION_state_id ::= ['INITIAL' | ... | 'FINAL']
>   STATION_PARTOF_opt ::= ['PARTOF' STATION_id ';']

STATION_attribute ::=

    [STATION_proctype |
    STATION_workstore |
    STATION_bus |
    STATION_device_list]

STATION_proctype ::= 'PROCTYPE' ':' processor_id [speed_descr] ';'
processor_id ::= id
speed_descr ::= 'AT' Int_const_denot 'MHz'
STATION_workstore ::= 'WORKSTORE' ':' workstore_descr ';'
workstore_descr ::= workstore_size_descr [workstore_space_descr]
workstore_size_descr ::= 'SIZE' int_const_denot workstore_size_denot
workstore_size_denot: ['B' | 'KB' | 'MB']
workstore_space_descr ::= 'SPACE' space_division space_access_attr

    space_access_speed
    {[space_divisions space_access_attr space_access_speed | ',']}

space_division ::= workstore_spc '-' workstore_spc
space_access_attr ::= ['READONLY' | 'DUALPORT']
space_access_speed ::= ['WAITCYCLES' int_const_denot

    | 'ACCESSTIME' float_const_denot 'SEC']

workstore_spc ::= int_const_denot | '"' id 'B' base_denot
base_denot ::= '1' | '2' | '3' | '4'
STATION_bus ::= 'BUS' Int_const_denot 'BIT' ';'
STATION_device_list ::= device_dscr {[device_dscr | ';']} ';'
device_dscr ::= [interface_descr | custom_device_descr]
interface_descr ::= 'INTERFACE' device_id {[device_descr_options]}
custom_device_dscr ::= 'DEVICE' device_id {[device_descr_options]}
device_ descr_options ::= 'DRIVER' ':' driver_id ';' |

    'ADDRESS' ':' workstore_spc ';' |
    'CONTROL' ':' workstore_spc ';' |
    'TRANSFER' ':' transfer_mode_denot ';' |
    'TRANSFERRATE' ':' transfer_rate ';' |

transfer_mode_denot ::= ['PACKAGE' int_const_denot 'BITS' | 'DMA']
device_id ::= id
STATION_state_id ::= 'STATEID' ':' '(' state_def {[state_def | ',']} ')' ';'
state_def ::= state_id ':' state_register_descr

    [':' state_register_dontcare_descr]

state_id ::= id
state_register_descr ::= BIT_string
state_register_dontcare_descr ::= BIT_string

The state register description must not be a BIT_string. Its state can be checked or changed using the following CM system calls GETSTATE and SETSTATE.

STATION_type_descr_opt::= STATION_type { type_specific_attrs }
STATION_type ::= 'STATIONTYPE' ':' station_type_denot ';'
station_type_denot ::= ['BASIC' |

> 'TASK' |
> 'KERNEL' |
> 'COMPOSITE']

type_specific_attrs ::= [task_processor_attrs | kernel_processor_attrs |]
task_processor_attrs ::= supervisor_spec
supervisor_spec ::= 'SUPERVISOR' ':' STATION_id ';'
kernel_processor_attrs ::= { [scheduling_spc |

> max_proc_spc |
> max_sema_spc |
> max_sign_spc |
> max_event_spc |
> max_event_queue_spc |
> mba |
> max_schedules_spc |
> RTC_base_spc] }

scheduling_spc ::= 'SCHEDULING' ':' scheduling_denot ';'
scheduling_denot ::= ['EDF' | 'MLF' | 'DMS' |

> 'RM' '(' FIXED_const_denot ')' | 'RR']

max_proc_spc ::= 'MAXPROC' ':' FIXED_const_denot ';'
max_sema_spc ::= 'MAXSEMA' ':' FIXED_const_denot ';'
max_sign_spc ::= 'MAXSIGN' ':' FIXED_const_denot ';'
max_event_spc ::= 'MAXEVENT' ':' FIXED_const_denot ';'
max_event_queue_spc ::= 'MAXEVENTQ' ':' FIXED_const_denot ';'
max_schedules_spc ::= 'MAXSCHED' ':' FIXED_const_denot ';'
mba ::= 'MBA' ':' FIXED_const_denot ';'
RTC_base_spc ::= 'TICK' ':' FLOAT_const_denot 'SEC' ';'

A.2   Net Division

In a net division the physical connections between stations are given by listing the point-to-point connections between their interfaces.

NET_division ::= 'NET' ':' { connection_spc } 'NETEND' ';'
connection_spc ::= endpoint_ety direction_qualifier endpoint_ety ';'

direction_qualifier ::= ['IN' | 'OUT' | 'INOUT']
endpoint_ety ::= [[endpoint_element | '+'] |]
endpoint_element ::= [user_id ':' { user_id ':' }

     [SYSTEM_id] qual_ety | SYSTEM_id qual_ety]

user_id ::= id ['(' enclosure ')']
SYSTEM_id ::= [STATION_id ','] local_SYSTEM_id
local_SYSTEM_id ::= [id ['(' enclosure ')'] |

     id '(' element_number ')']

enclosure ::= first_element ':' last_element
element_number ::= Int_const_denot
first_element ::= Int_const_denot
last_element ::= Int_const_denot
qual_ety ::= ['*' connection_point { '*' connection_point }

     [',' Int_const_denot] |]

connection_point ::= [id | Int_const_denot |

     '(' enclosure ')' ['/' Int_const_denot]]

A.3   System Division

A system division encapsulates the hardware architecture description and the assignment of symbolic names to hardware devices. The components as described in station and net divisions are used.

SYSTEM_division ::=

     'SYSTEM' ':'
     { STATION_SYSTEM_division }
     'SYSEND' ';'

STATION_SYSTEM_division ::=

     STATION_names
     { device_or_connection_spc ';' }

device_or_connection_spc ::=

     [device_spc | connection_spc]

device_spc ::= user_id ':' { user_id ':' }

     SYSTEM_id

A.4   Collection Division

A configuration division is dealing with a SW architecture. The largest program component that is associated with a station and its state is a "collection" of "modules". Modules consist of "tasks", which may communicate through "PORTS" and their IMPORT/EXPORT structures. Each SW component has its unique name for reference. Modules are further described by their IMPORT and EXPORT parts, in which it is stated, which data structures and task references are shared with (exported to) other modules.

Tasks are described by their scheduling parameters. Two attributes can be assigned to tasks. The INIT attribute is associated with the initial task. The KEEP attribute is associated with a critical task.

Collections of modules are loaded to stations. It is also possible to specify under which conditions certain collections are to be removed from a station and which collections to be loaded instead (reconfiguration actions). These conditions are station-state-dependent and are being performed by the station's CM.

The connections between the ports are described by their directions and line attributes. Line attributes state which connections are always followed (VIA attribute), and which ones can be chosen from a list, based on the PREFER attribute.

CONFIGURATION_division ::=

    'CONFIGURATION' ':'
    configuration_stmts
    'CONFEND' ';'

configuration_stmts ::= initial_part [reconfiguration_part]
initial_part ::= COLLECTION_definition

    { COLLECTION_definition }
    initial_LOAD_stmt
    CONNECT_stmt
    { initial_LOAD_stmt CONNECT_stmt }

COLLECTION_definition ::=

    'COLLECTION'
    COLLECTION_id
    PORT_enumeration
    MODULE_definition
    { MODULE_definition }
    'COLEND' ';'

COLLECTION_id ::= id
PORT_enumeration ::= PORT_spc {[PORT_spc |]}
PORT_spc ::= 'PORT' PORT_id ':' PORT_attr ';'
PORT_id ::= id
PORT_attr ::= in_or_out_PORT_attr [PORT_signals]

in_or_out_PORT_attr ::= [in_PORT_attr I out_PORT_attr]
in_PORT_attr ::= 'IN' single_trf_item_type synch_mechanism
out_PORT_attr ::= 'OUT' single_trf_item_type [wait_option_one]
PORT_signals ::= 'SIGNAL' '(' SIGNAL_id {[SIGNAL_id I ',']} ')'
SIGNAL_id ::= id
single_trf_item_type ::= [basic_type I struct_dcl_attr]
synch_mechanism ::= [buffer_option I reply_option_one]
wait_option_one ::= 'WAIT' [single_trf_item_type]
buffer_option ::= ['BUFFER' '(' FIXED_const_denot ')' I]
reply_option_one ::= 'REPLY' [single_trf_item_type]
MODULE_definition ::=

     'MODULE' MODULE_id
     'IMPORTS' id_list_pack ';'
     'EXPORTS' id_list_pack ';'
     TASK_definition { TASK_definition }
     'MODEND' ';'

MODULE_id ::= id
id_list_pack ::= id { [id I ','] }
TASK_definition ::=

     'TASK' TASK_id ['INIT' I 'KEEP']
     [priority_spc I responsetime_spc I]
     'TASKEND' ';'

TASK_id ::= id
priority_spc ::= ['PRIO' I 'PRIORITY'] Int_const_denot
responsetime_spc ::= ['RESPT' I 'RESPONSETIME'] dur_exp
initial_LOAD_stmt ::= LOAD_phrase ';'
LOAD_phrase ::=

     'LOAD' COLLECTION_id
     {[COLLECTION_id I ',']}
     'TO' STATION_id

CONNECT_stmt ::=

     'CONNECT' global_PORT_list
     PORT_direction_qualifier
     global_PORT_list [line_map] ';'

global_PORT_list ::= global_PORT_id {[global_PORT_id I ',']}
global_PORT_id ::= COLLECTION_id '.' PORT_id
PORT_direction_qualifier ::= ['<-' I '->' I '<->']
line_map ::= [fixed_line I preferred_line]
fixed_line ::= 'VIA' line_list
preferred_line ::= 'PREFER' line_list
line_list ::= user_id {[user_id I ',']}

reconfiguration_part ::= reconfiguration_stmt { reconfiguration_stmt }
reconfiguration_stmt ::=

     'STATE' '(' state_expr ')' reconfiguration_block ';'

state_expr ::= [state_denotation |

     not_operator state_denotation |
     state_denotation and_operator state_expr]

state_denotation ::= STATION_id '.' state_id
state_id ::= id
not_operator ::= ['NOT' | '-']
and_operator ::= ['AND' | '&']
reconfiguration_block ::= 'BEGIN' reconfiguration_actions 'END'
reconfiguration_actions ::=

     { DISCONNECT_stmt }
     { REMOVE_stmt }
     { reconfiguration_LOAD_stmt }
     { CONNECT_stmt }

DISCONNECT_stmt ::=

     'DISCONNECT' global_PORT_list
     PORT_direction_qualifier
     global_PORT_list ';'

REMOVE_stmt ::=

     'REMOVE' COLLECTION_id
     {[COLLECTION_id | ',']}
     'FROM' STATION_id ';'

reconfiguration_LOAD_stmt ::= LOAD_phrase [RESIDENT_option] ';'
RESIDENT_option ::= 'RESIDENT'

Appendix B
Graphical Architecture Description

The architecture name is identical to project name. The architecture may be attributed SIL and SL levels according to the textual specification. The architecture description is composed of hardware and software specifications.

A hardware model consists of STATIONs, being the processing nodes of a system. Their components (see Fig. B.1) are chosen from a list of general components such as processors, memories or interfaces. These components determine the structure of stations and, on the other hand, also represent their resources and provide the necessary timing information for schedulability analysis.

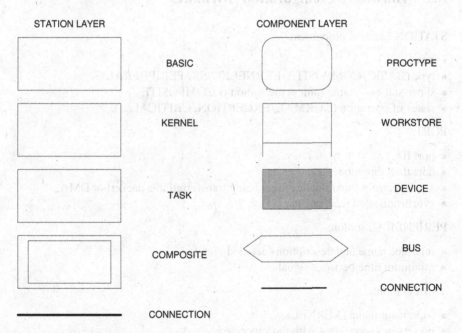

Fig. B.1 Hardware architecture constructs of Specification PEARL

© Springer International Publishing Switzerland 2016
R. Gumzej, *Engineering Safe and Secure Cyber-Physical Systems*,
Studies in Computational Intelligence 632, DOI 10.1007/978-3-319-28905-2

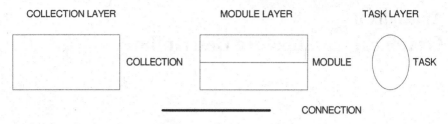

Fig. B.2 Software architecture constructs of Specification PEARL

A software model is composed of COLLECTIONs, which are mapped to the STATIONs of a hardware model, depending on their state information (see Fig. B.2). They consist of MODULEs of TASKs. At any time, there is exactly one collection assigned to run on a station. Thus, the collection is also the unit of dynamic reconfiguration.

STATIONs and COLLECTIONs (station and collection layers) and their CONNECTIONS (between ports on both layers) have been identified to be the basic building blocks of distributed systems. Their components, extended with attributes, can present the same semantics as their textual counterpart description.

B.1 Hardware Configuration Attributes

STATION (general properties):

- name,
- type (BASIC, COMPOSITE, KERNEL, TASK, PERIPHERAL),
- super-station—name (unless the station is COMPOSITE),
- states of operation (NORMAL, EXCEPTION, CRITICAL,...).

PORT:

- port ID,
- data flow direction,
- single transfer unit (smallest item being transferred in a packet) or DMA,
- synchronisation mechanism.

PERIPHERAL station:

- interface name and description (detailed view),
- minimum time between signals.

TASK station:

- supervisor name (KERNEL),
- port name (connection with the supervisor).

KERNEL station:

- real-time clock resolution,
- scheduling strategy,
- MAXTASK,
- MAXSEMA,
- MAXEVENT,
- MAXQEVENT,
- MAXSCHEDEVENT.

PROCTYPE:

- processor ID,
- processor speed.

WORKSTORE:

- memory area size,
- access type (READ/WRITE, READ, EXECUTE),
- number of wait cycles to access the area.

DEVICE:

- Standard devices are identified only by their identifiers, whereas their behaviour is known.

  INTERFACE:

  - interface ID,
  - driver ID and start address of the driver,
  - data transfer direction,
  - transfer speed,
  - one package size or DMA,
  - interrupt vector and level.

  TIMER:

  - Timer ID (driver or device),
  - timer activation time, period between signals and duration of its activity,
  - timer resolution.

  SHARED STORAGE:

  - shared storage ID or address,
  - signal trigger condition (value change or logical condition),
  - comparison register address.

B.2   Software Configuration Attributes

COLLECTION:

- station ID (residence),
- assignment of logical names to connections between collections (tasks) and stations.

MODULE:

- module ID,
- collection ID,
- properties and methods, which the model exports/imports to/from another modules.

TASK:

- task ID,
- module ID,
- trigger condition (on demand, timer, interrupt, signal),
- deadline,
- alternative task ID (scheduled instead if schedule becomes infeasible).

Appendix C
CM API

The application programming interface of the CM has the following functions:
(Re-) Configuration:

$Cm\_Init(S)$ —to initialise the station S and load the initial software configuration, and

$Cm\_Reset(S)$ —to restart the station with the initial software/hardware configuration.

Station state monitoring:

$Cm\_Getstate(S)$ —to retrieve the current state of station S, and
$Cm\_Setstate(S, state)$ —to change the current state of station S to "state".

Inter-station communication:

$Cm\_Transmit(TCBi, portID, msg\_buff[])$ —message transmission via a connection,

$Cm\_Reply(TCBi, portID, msg\_buff[])$ —response message transmission via a connection, and

$Cm\_Receive(TCBi, portID, msg\_buff[])$ —message receipt via a connection, where TCBi denotes the index of the task's control block (TCB), portID the name of the port, and msg_buff[] the buffer for the message.

The connections are established through ports of the software architecture and associated devices of the hardware architecture. The attributes of ports represent the communication parameters (smallest package, protocol, etc.) and routing parameters (VIA/PREFER). Routing affects the way the hardware communication devices are used. The attribute VIA determines the exact line to be used, while PREFER is usually assigned to the most trusted line in a list. Lines represent connections between hardware architecture devices (e.g. interfaces).

In asymmetrical architectures, direct calls to real-time operating system functions are not always possible. Hence, substitute RTOS API functions are called to generate appropriate system request messages to the CM of the RTOS's processing (supervisor) node. The parameters of such system requests are extracted from the

© Springer International Publishing Switzerland 2016

R. Gumzej, *Engineering Safe and Secure Cyber-Physical Systems*,
Studies in Computational Intelligence 632, DOI 10.1007/978-3-319-28905-2

transferred messages in concordance with a predefined coding scheme also used in the construction of the parameter set.

To enable uniform handling of system requests, the RTOS API has been designed in a way enabling the transformation of system calls to parameter strings, which can be routed to the RTOS interface procedure directly, or sent to the KERNEL station for handling. Two additional internal functions have been introduced in the CM interface for this reason:

$Cm\_SysRequest(S, sys\_par[])$ —send system call parameters for processing to the RTOS, and

$Cm\_SysResult(S)$ —get result from the RTOS (store the result of the system call and process a possible context switch request for the local dispatcher routine of the CM ($CM\_System(S)$).

Appendix D
RTOS API

Here the system calls of the HaRTOS operating system are described. They are introduced in the way they support the real-time behaviour of application programs written in PEARL. For each system call, first a description of the language constructs which require operating system services and generate adequate calls (schedule management, tasking operations, inter-task synchronisation, etc.) is given. This description is followed by the specification of its implementation.

D.1 Task Scheduling

Task activation and continuation operations can be scheduled to be executed upon "fulfillment of a condition" or a combination thereof, called event(s). These conditions can be time-related or not. In case of a time schedule, the parameters being transmitted to the RTOS kernel consist of the start time, period and validity duration of the schedule. An event is triggered upon reaching the start time and repeated after the specified period has elapsed during the specified duration. A non-temporal condition can denote an internal or an external signal (interrupt), being fulfilled only once at a time. Combinations of temporal and non-temporal schedules are also possible, in which case the condition, being fulfilled first, triggers the start of the operation.

Syntax:

Start_condition ::=

    'AT' Expression$Time [Frequency] |
    'AFTER' Expression$Duration [Frequency] |
    'WHEN' Name$Interrupt [AFTER Expression$Duration] [Frequency] |
    Frequency

Frequency ::=

    { 'ALL' | 'EVERY' } Expression$Duration
    [{ 'UNTIL' Expression$Time } | { 'DURING' Expression$Duration }]

© Springer International Publishing Switzerland 2016
R. Gumzej, *Engineering Safe and Secure Cyber-Physical Systems*,
Studies in Computational Intelligence 632, DOI 10.1007/978-3-319-28905-2

A schedule option can be associated with the activation, continuation and resumption operations. In case the appropriate parameters for a specific schedule are present, a schedule is set up for the operation to be executed upon fulfillment of the scheduling condition.

D.1.1   Task Activation

To execute a task, it needs to be activated. As for all tasking operations, this operation's key parameter is the task identification, which contains the index of the task's entry in the TCB table. This entry comprises data required by the kernel routines for scheduling. If priority scheduling is applied, the obligatory parameter of a task is its priority. In addition, if a deadline is to be assigned to the task being activated, a response time needs to be specified. In this case the task's deadline is calculated as the sum of the real-time clock's current reading and the response time.

Syntax:

Task-Activation ::=

    [Start_condition]
    'ACTIVATE' Name$Task
    [Priority-Clause | Response-time-Clause] ';'

   Priority-Clause ::= { 'PRIO' | 'PRIORITY' } Expression$positive-integer
  Response-time-Clause ::= { 'RESPT' | 'RESPONSETIME' } Expression$Duration

System call:

Pearl_Task_Activate (TCB_id, [prio | rest]);

Description:

If the current number of task activations exceeds the maximum number specified, the operation is not executed. If the task is already active, the number of buffered activations is increased and the circular buffer of buffered activations is updated with the new activation information. Any exhausted schedules of the task are removed. If the task is not active, it is activated.

D.1.2   Task Termination

An active task can be terminated, i.e. made inactive. If the task is running, its execution is immediately stopped, its entry is deleted from the "ready"-list and its schedules are deleted. In case a buffered activation exists for this task, it is inserted into the "ready"-list. Synchronisers seized by the task are released.

Syntax:

Task-Termination ::= 'TERMINATE' [Name$Task] ';'

System call:

Pearl_Task_Terminate (TCB_id);

Description:

The operation is not executed if the task is non-active. The task is pre-empted if it is in the "run" state. It is removed from the ready-queue, if it is in the "ready" state. Consequently, the tasks in the queue are rescheduled by the RTOS. Any remaining schedules for the task are removed and any synchronisers seized by the task are released. The task's circular activation buffer is updated and the number of activations is decreased. If a buffered activation exists for the task, it is scheduled for execution.

D.1.3   Task Prevention

Reactivations of active tasks can be prevented by deleting their schedules and the buffered task activations.

Syntax:

Task-Prevention ::= 'PREVENT' [Name$Task] ;

System call:

Pearl_Task_Prevent (TCB_id);

Description:

The buffered activations of the task are disabled by resetting the appropriate indexes and counter. The task's schedules are removed.

D.1.4   Task Suspension

An active task can be suspended. If the task is running, it is immediately pre-empted. The task is removed from the "ready"-list. The exhausted schedules, which have been set up for this operation, are deleted.

Syntax:

Task-Suspension ::= 'SUSPEND' [Name$Task] ';'

System call:

Pearl_Task_Suspend (TCB_id);

Description:

If the task is non-active, the operation is not executed. If it is in the state "run", it is pre-empted. If the task is in the state "ready", it is removed from the ready-queue and the tasks in the queue are rescheduled.

D.1.5   Task Continuation and Resumption

This operation enables the continuation of a suspended task. A new deadline for the task process is set up, and the task is inserted (if it is not suspended for synchronisation) into the "ready"-list. The exhausted schedules, which have been set up for this operation, are deleted. If the deadline for the task is to be updated, a response time needs to be specified. Hence, the deadline is calculated as the sum of the real-time clock's current reading and the response time given.

Syntax:

Task-Continuation ::=

> ['AT' Expression$Time |
> 'AFTER' Expression $Duration |
> 'WHEN' Name$Interrupt]
> 'CONTINUE' [Name$Task]
> [Priority-Clause | Responsetime-Clause] ';'

System call:

Pearl_Task_Continue (TCB_id, [prio | rest]);

Description:
If the task is to be scheduled for continuation, the parameters of the schedule are input into the schedule. Any remaining schedules for the task are removed. If the task is not "suspended", the operation is not executed. Based on the current time and the response time specified, a new deadline is calculated for the task. In case only a priority is given, this new priority is assigned to the task. If the task is not suspended for synchronisation, it is scheduled for execution.

If a task is to be suspended immediately and a schedule for its CONTINUATION to be set up, a single PEARL statement (RESUME) can be employed. In this case the schedule parameters are obligatory.

Syntax:

Task-Resumption ::=

> { 'AT' Expression$Time |
> 'AFTER' Expression $Duration |
> 'WHEN' Name$Interrupt }
> 'RESUME' ';'

System call:

Pearl_Task_Resume (TCB_id);

Description:
The task suspends itself and schedules its continuation for the fulfillment of the schedule specified.

D.1.6   Normal Task End

Upon reaching a task's normal end, this operation is executed. It removes the task from the "ready"-list and, if present, inserts a buffered task activation into this list. If any synchronisers were seized during the task's execution, they are released.
   Syntax:

   Task-End ::= 'END' ';'

System call:

   Pearl_Task_End (TCB_id);

   Description:
   If the task is in the "run" state, it is pre-empted. The task is removed from the "ready"-queue and the other tasks there are rescheduled. The circular activation buffer is updated and the activation counter decremented.

D.1.7   Synchronisation Constructs and Critical Regions

Critical regions are parts of the tasks' application program code, which perform some operation on shared objects and must, therefore, be synchronised with parts of other tasks' application program code using the same objects, in order not to corrupt each others' data.
   Syntax:

   Task-Synchronisation ::=

           { 'TRY' | 'REQUEST' | 'RELEASE' }
           Name$Sema [, Name$Sema]...
           ['TIMEOUT' Time$Cond] ';'

System calls:

   Pearl_Sema_Try (TCB_id, Sema_id);
   Pearl_Sema_Request (TCB_id, Sema_id, to);
   Pearl_Sema_Release (TCB_id, Sema_id);

   Description:
   The statements TRY and REQUEST execute the system call Lock. If the lock operation fails on a given semaphore, the corresponding task is suspended for synchronisation and placed in a queue until the semaphore is RELEASEd. If the timeout option is selected, the task is woken up when it elapses. The statement RELEASE is executed by the system call Unlock. Upon release of the semaphore its queue is checked and an appropriate task is continued.
   When the execution of a task's application code reaches a "REQUEST", the parameters of the associated system call are sent to the kernel. TRY attempts to seize the

synchroniser with the identification number given by "Sema_id". If the synchroniser is free, it is locked by REQUEST and the application is notified that it can continue executing. When the execution reaches "RELEASE", the parameters of the pertaining system call are sent to the kernel. RELEASE frees the synchroniser seized by REQUEST. In case the REQUEST system call fails to seize the synchroniser, the task is suspended and inserted into the synchroniser's queue. After the release of the synchroniser, its queue is checked and a waiting request processed, which results in the continuation of a suspended task.

D.2   SYSTEM DIVISION Constructs

The system division of an application contains declarations of system variables, which have global scope within the application. The constructs described here are all declared within applications and are initialised by means of system calls, because they are managed by the kernel. In Specification PEARL, the SYSTEM DIVISION is extended by constructs describing the stations and their interconnections.

The interrupts declared have to be enabled by the kernel. The declaration of a synchroniser produces its "Sema_id". To each synchroniser an enter-range (i.e. the number of requests, which may seize it at the same time) is assigned. The signals used later on in the program also have to be declared here.

Syntax:

Systempart ::=

    'SYSTEM' ';' [Username-Assignment$for-Dation-Interrupt-or-Signal]

Username-Assignment ::=

    Identifier$Username: Identifier$System-name [(nngz$Index)] [* nngz$Channel [
      * nngz$Position] [, nngz$Width]] ';' |
    Identifier$Username : Identifier$SIGNAL-Systemname [(Identifier$Error-number
      [,Identifier$ Error-number]...)] ';'

    nngz ::= integer-without-precision$non-negative

These system variables are declared and initialised in the CM/RTOS initialisation file.

D.2.1   State Acquisition Constructs (CM)

In real-time programming it is often useful for application programmers to be able to observe the values of some system variables. This way some system-level decisions can be taken at the application level.

Syntax:

> Now-Statement ::= 'NOW' ';'
> Getstate-Statement ::= 'GETSTATE' ';'
> Setstate-Statement ::= 'SETSTATE' State$Identifier ';'

System calls:

> Pearl_Rtc_Now();
> Cm_Getstate();
> Cm_Setstate(Tstate state);

Description:
NOW returns the current value of a station's real-time clock. The NOW, GETSTATE and SETSTATE statements are system calls to the station's CM.

D.2.2 Enabling/Disabling of External Events

The following statements handle external events.
Syntax:

> Interrupt-Statement ::=
> { 'ENABLE' | 'DISABLE' | 'TRIGGER' } Name$Interrupt ';'

System calls:

> Pearl_Int_Enable(int);
> Pearl_Int_Disable(int);
> Pearl_Int_Trigger(int);

Description:
The ENABLE, DISABLE and TRIGGER statements represent system calls to the station's CM.

D.2.3 Signal Handling

The following statements handle internal events (signals). With the ON-statement the execution of a task is scheduled to react on the occurrence of a signal. INDUCE-statement triggers the signal.
Syntax:

> SchedulingSignalReaction ::=
>
> > 'ON' Name§Signal { ['RST' (Name§ErrorVariable-FIXED)] ':'
> > SignalReaction | 'RST' (Name§ErrorVariable-FIXED) } ';'
> > SignalReaction ::=
> > UnlabeledStatement

and
InduceStatement ::=
'INDUCE' Name§Signal ['RST' (Expression§ErrorNumber)] ';'

System call:

Pearl_Sign_Induce (Sign_id);

Description:
The INDUCE statement represents a system call to the station's CM. Scheduling a task for the occurence of a signal is the same as scheduling it to an occurence of an interrupt—they both represent non-temporal schedules. The main difference is in the fact that not only the task execution is triggered, but also an optional paramater is passed to the task, allowing it to handle the event properly.

D.2.4   PORT I/O

The following statements are used for PORT-TO-PORT communication both between and within tasks and collections.
Syntax:

Transmit-Statement ::= 'TRANSMIT' transmit_params ';'
transmit_params ::= trf_expr 'TO' PORT_id [wait_option_two]
trf_expr ::= exp
wait_option_two ::= 'WAIT' [trf_expr] [timeout_part]
timeout_part ::= 'TIMEOUT' duration_expr ['REACT' unlabelled_stmt]
Receive-Statement ::= [simple_receive | selective_receive]
simple_receive ::= 'RECEIVE' receive_params [end_receive_part]
receive_params ::= trf_expr 'FROM' PORT_id [reply_option_two]
reply_option_two ::= ['REPLY' | [unlabelled_stmt] 'REPLY' trf_expr]
end_receive_part ::= [otherwise_part | timeout_part]
otherwise_part ::= 'OTHERWISE' unlabelled_stmt
selective_receive ::=

        'RECEIVE' 'SELECT' receive_params { receive_params }
        'OR' receive_params { receive_params }
        {'OR' receive_params { receive_params }}
        [end_receive_option]

    end_receive_option ::= end_receive_part

System calls:

Cm_Transmit(Task_id, Port_id, Transmit_par, Msg);
Cm_Receive(Task_id, Port_id, Receive_par);
Cm_Reply(Task_id, Port_id, Msg);

Description:
The TRANSMIT, RECEIVE and REPLY statements are system calls to the station's CM.

Appendix E
Project Layout

Configuration files for the designed system model are created automatically from the Specification PEARL CASE environment. They are laid out in the sequel. The first part is devoted to hardware (HW) architecture configuration files, which are meant for target platform compilation, while the second—the software (SW) architecture configuration files—are meant for both—target platform and simulation purposes—and are parameterised accordingly. The CM/RTOS libraries are integrated with the system model and parameterised accordingly. The names in angle brackets represent placeholders for components' actual names. Hence the listed files are to be considered templates of the actual project configuration files.

E.1   HW Architecture

Station definition is done in <Station> header and source files. The header file comprises station attributes and establishes external dependencies. The platform-specific libraries on the station's components are included only if the simulation switch ("SIMULATION") is not set.

```
/* ------------------ <Station>.h ------------------ */

#define <Station>_PARTOF <super_station_id>
#define <Station>_TYPE <station_type>

// type-dependent information (TASK_STATION)
#define <Station>_SUPERVISOR_ID <kernel_station_id>

// type-dependent information (KERNEL_STATION)
#define MAX_TP <max_tp>
#define SCHED_ID <scheduling_id>
#define MAX_PRIO <priority_levels>
#define MAX_TASK <max_tasks>
#define MAX_SYN <max_sema>
#define MAX_SIG <max_signal>
```

© Springer International Publishing Switzerland 2016
R. Gumzej, *Engineering Safe and Secure Cyber-Physical Systems*,
Studies in Computational Intelligence 632, DOI 10.1007/978-3-319-28905-2

```
#define MAX_INT <max_int>
#define MAX_QEV <max_queued_events>
#define MAX_SCH <max_schedules>
#define RTC_RES <RTC_resolution>
#define MBA <max_buff_act>
#define MAX_EVENT 1+ MAX_SIG+ MAX_INT

// states enumeration
#define UNDEFINED_STATE −1
#define INITIAL_STATE 0
#define MONITORING_STATE 1
#define ANALYSIS_STATE 2
#define PLANNING_STATE 3
#define EXECUTION_STATE 4
// any other (e.g. exception) state definitions

#ifndef SIMULATION
#include "<Station>_Proctype.h"
#include "<Station>_Workstore.h"
#include "<Station>_<Device>.h"
#endif

extern void <Station>_reconfigure(char s1, char s2);
```

The source file implements the station definition and operation procedure. The method <Station>_reconfigure(s1, s2) is responsible for loading a collection to the station for execution and establishing its connections according to station state.

```
/* ----------------- <Station>.c ----------------- */

#include "<architecture>.h"
#include "<Station>.h"

void <Station>_reconfigure(char s1, char s2)
{
if (s1!=s2) {
      switch (s1) {
            // nothing to unload (initially)
               case UNDEFINED_STATE: break;
          // unload current collection
          case State1: {
               <collection_name1>_disconnect();
               <collection_name1>_unload();
          }
          // any subsequent variant
     }
        switch (s2) {
        // load new collection
        case State2 : {
             <collection_name2>_connect();
             <collection_name2>_load();
        }
        // any subsequent variant
     }
```

```
}
}
```

The <Station>_reconfigure(s1, s2) method also performs any subsequent reconfiguration operations on the station. In case of building the application for non-simulation environments, an additional file is built for the initialisation of execution at each station node:

```c
/* ------------------- <Station>_main.c ------------------- */

#include "<architecture>.h"
#include "cm.h"

#undef SIMULATION

void main() {
    int s=findStation("<Station name>");
    Cm_Init(s);
    Cm_Reset(s);
    while (1) Cm_System(s);
}
```

The header files on station components comprise their attributes. The processing units' header file template:

```c
/* ------------------- <Station>_Proctype.h ------------------- */

#define <Station>_PROCTYPE <processor_id>
#define <Station>_PROCSPEED <processor_speed>
```

The storage units' header file template:

```c
/* ------------------- <Station>_Workstore.h ------------------- */

#include <alloc.h>
// allocation unit, depending on bus width
typedef unsigned long int bword; // 32 bit
// any subsequent variant
typedef unsigned char bword; // 8 bit

bword <ws_name>[<ws_size>];
extern bword *<ws_name>;
```

Generic devices and interfaces header file templates:

```c
/* ------------------- <Station>_<Device>.h ------------------- */

#define <Station>_DEVICE_ID <device_id>
extern void <Station>_<device_id> ();


/* ------------------- <Station>_<Interface>.h ------------------- */

#define <Station>_DEVICE_ID <device_id>
#define CTRLREG <control_register>
#define DATAREG <data_register>
```

```
#define TRANSFER_TYPE <type_of_transferred_packet>
#define TRANSFER_RATE <packets_per_second>
extern void <Station>_<device_id>_<driver_id> ();
```

E.2   SW Architecture

The software configuration is established through collection, module and task header and source files' templates:

```
/* ----------------- <Collection>.h --------------- */

// Module enumeration
#include "<Module1>.h"
// any subsequent modules
#include "<ModuleN>.h"

// Port enumeration
extern port *<port_name>;

extern void <name>_load ();
extern void <name>_connect();
extern void <name>_unload();
extern void <name>_disconnect();
```

The collection is defined by the methods, which define how it is loaded and connected and unloaded and disconnected respectively. Its dependencies and port-connections are imported by its header file.

```
/* ----------------- <Collection>.c ---------------- */

#include "<architecture>.h"
#include "<Collection>.h"
// Port enumeration
port *<port_name>;

// (initial) LOAD statement
void <name>_load ()
{
    // port description
    <port_name>=new port(<buffer_size>);
    <port_name>->setDataDir(<port_direction>);
    <port_name>->setSyncMech(<port_sync_mech>);
    // if RTOS is present
    Pearl_Task_Activate(<task_name1>);
    Pearl_Task_Activate(<task_name2>);
    // any subsequent variant
    Pearl_Task_Activate(<task_nameN>);
    // otherwise set initial active Collection and Task manually
    st[si].aCollection=ci; ct[ct].aTask=ti;
}
```

```
// REMOVE statement
void <name>_unload ()
{
    // if RTOS is present
    Pearl_Task_Terminate(<task_name1>);
    Pearl_Task_Terminate(<task_name2>);
    // any subsequent variant
    Pearl_Task_Terminate(<task_nameN>);
}

// CONNECT statement
void <name>_connect()
{
    // net division
    Line *<l>_<port>=new Line(<port_name1>,<port_name2>,<line_attr>);
    <port_name1>-->addLine(<l>_<port>);
    // any subsequent variant
}

// DISCONNECT statement
void <name>_disconnect()
{
    <port_name1>-->removeLine(<l1>_<port>);
    // any subsequent variant
    <port_nameN>--> removeLine(<lN>_<port>);
}
```

The load() and connect() methods mentioned here perform the initialisation of collections and establishment of their global inter-connections. The unload() and disconnect() methods are meant for any subsequent reconfiguration operations.

Module header and source files establish libraries of tasks that can communicate through exported task definitions and external variables.

```
/* ----------------- <Module>.h ----------------- */
// Task enumeration
#include "<Task1>.h"
// any subsequent tasks
#include "<TaskN>.h"

extern <var>; // shared storage / structures only

/* ----------------- <Module>.c ----------------- */

#include "<Module>.h"

<var>; // shared storage (variables / structures) only
```

Task header and source files define program tasks as derived from task TSTD representations.

```
/* ----------------- <Task>.h ----------------- */

#ifdef SIMULATION
#include "SimulationTypes.h"
```

```
//  main task program
extern void <task_name>_main(TSimulationUnit*, int &);
#else
//  main task program
extern void <task_name>_main(int &);
#endif
```

```
/* ------------------ <Task>.c ------------------ */
```

```
#include "cm.h"
//  if RTOS is enabled
#include "rtos.h"
#include "<architecture>.h"
#include "<Collection>.h"
#include "<Module>.h"
#include "<Task>.h"
```

```
//  TASK description
#ifdef SIMULATION
void <task_name>_main(TsimulacijskaEnota*e, int &s) {
    //  task TSTD representation
}
#else
void <task_name>_main(int &s) {
    //  task TSTD representation
#endif
}
```

Architecture header and source files combine all components of the architecture configuration definition.

```
/* ------------------ <architecture>.h ------------------ */
```

```
#include "port.h"
```

```
/* identifiers are up to 20 characters long */
#define MAX_NAME 20
```

```
/* types of stations (device encoding) */
#define BASIC 50
#define TASK 51
#define KERNEL 52
#define COMPOSITE 53
```

```
/* SIL conformity (default 1) */
#define SIL 1
```

```
/* SL conformity (default 1) */
#define SL 1
```

```
/********* STATIONs **********/
#include "<Station>.h"
//  any subsequent stations
```

```
/********** COLLECTIONS ********/
#include "<Collection>.h"
// any subsequent collections

/* resources parameterisation (station's maximum/cumulative resources) */
#define MAX_STATION 1
#define MAX_COLLECTION 1
#define MAX_TASK 1
#define MAX_INT 1
#define MAX_SYN 1
#define MAX_SIG 1
#define MAX_SCHED 2
#define MAX_EVENT (1+ MAX_SIG+ MAX_INT)
#define MAX_PORT 1

#define SIMULATION
struct SStationTable {
    char sname[MAX_NAME];
    char type;
    char sport[MAX_NAME];
    char sreg;
    int aCollection;
    void (*reconfigure)(char,char);
};

struct SCollectionTable {
    char StationIdx;
    char cname[MAX_NAME];
    int aTask;
    char sts;
};

struct STaskTable {
    char StationIdx;
    char cname[MAX_NAME];
    char mname[MAX_NAME];
    char tname[MAX_NAME];
    bool keep;
    int schp;
    int result;
    int state;
    void (*func) (TSimulationUnit*, int&);
};

struct SSemaTable {
    char cname[MAX_NAME];
    char sname[MAX_NAME];
    int enter_range;
};

struct SSignalTable {
    char cname[MAX_NAME];
    char sname[MAX_NAME];
```

```
};

struct SIntTable {
    char StationIdx;
    char iname[MAX_NAME];
};

struct SPortTable {
    char cname[MAX_NAME];
    char pname[MAX_NAME];
    port *p;
};

extern SSignalTable sigt[MAX_SIG];
extern SScmaTable semt[MAX_SYN];
extern SPortTable pt[MAX_PORT];
extern STaskTable tt[MAX_TASK];
extern SCollectionTable ct[MAX_COLLECTION];
extern SStationTable st[MAX_STATION];
extern int findStation (char *);
extern int findCollection (char *);
extern int findTask (char *,char *,char *);
extern int findSema(char *);
extern int findSignal(char *);
extern port* findPort(char *);

/* ----------------- <architecture>.c ----------------- */

#include <string.h>
#include "<architecture>.h"

int findStation(char *name)
{
    for (int i=0;i<MAX_STATION; i++) {
        if (strcmp(name,st[i].sname)==0)
            return i;
    }
    return -1;
}

int findCollection(char *name)
{
    for (int i=0;i<MAX_COLLECTION; i++) {
        if (strcmp(name,ct[i].cname)==0) return i;
    }
    return -1;
}

int findTask(char *cname, char* mname, char *name)
{
    for (int i=0;i<MAX_TASK; i++) {
        if ((strcmp(cname,tt[i].cname)==0) &&
            (strcmp(mname,tt[i].mname)==0) &&
            (strcmp(name,tt[i].tname)==0))
```

```
            return i;
    }
    return −1;
}

int findSignal(char *name) {
    char gname[2*MAX_NAME+1];
    for (int i=0;i<MAX_SIG; i++) {
        strcpy(gname,sigt[i].cname);
        strcat(gname,"_");
        strcat(gname,sigt[i].sname);
        if (strcmp(name,gname)==0)
            return i;
    }
    return −1;
}

int findSema(char *name) {
    char gname[2*MAX_NAME+1];
    for (int i=0;i<MAX_SYN; i++) {
        strcpy(gname,semt[i].cname); strcat(gname,"_");
        strcat(gname,semt[i].sname);
        if (strcmp(name,gname)==0)
            return i;
    }
    return −1;
}

port* findPort(char *name) {
    char gname[2*MAX_NAME+1];
    for (int i=0;i<MAX_PORT; i++) {
        strcpy(gname,pt[i].cname);
        strcat(gname,"_");
        strcat(gname,pt[i].pname);
        if (strcmp(name,gname)==0)
            return pt[i].p;
    }
    return NULL;
}

SStationTable st[MAX_STATION]={{}};
SCollectionTable ct[MAX_COLLECTION]={{}};
STaskTable tt[MAX_TASK]={{}};
SSignalTable sigt[MAX_SIG]={{}};
SSemaTable semt[MAX_SYN]={{}};
SIntTable it[MAX_INT]={{}};
SPortTable pt[MAX_PORT]={{}};
```

The above-mentioned configuration files are combined with *cm.h, rtos.h* and associated libraries into a whole when compiling the design for execution/simulation.

```
/* ─────────────────── cm.h ─────────────────── */

/* Configuration management interface: */
```

```
extern __declspec(dllexport) void Cm_Init(char);
extern __declspec(dllexport) void Cm_Reset(char);
extern __declspec(dllexport) char Cm_Getstate(char);
extern __declspec(dllexport) void Cm_Setstate(char, char);
extern __declspec(dllexport) char Cm_Receive(int,char*, char*);
extern __declspec(dllexport) char Cm_Transmit(int,char*, char*);
extern __declspec(dllexport) char Cm_Reply(int,char*, char*);
extern __declspec(dllexport) void Cm_SysRequest(char, char*);
extern __declspec(dllexport) void Cm_SysResult(char);
extern __declspec(dllexport) void Cm_System(char);
```

The station's configuration manager needs to be aware of the architecture specification in order to manage the architecture accordingly. Hence, the architecture configuration is included.

```
/* ----------------- cm.c ----------------- */

#include <string.h>
#include "<architecture>.h"
#include "cm.h"
#include "port.h"
#include "rtos.h"

/*--------------- Configuration management: --------------- */

/* Station initialisation and initial loading */
void Cm_Init(char S)
{
    // initial station operational state
    st[S].sreg = UNDEFINED_STATE;
    // RTOS init
    if (st[S].type==KERNEL_STATION) RTOS_Init();
}

/* Station reinitialisation and initial loading */
void Cm_Reset(char S)
{
    st[S].reconfigure(st[S].sreg,INITIAL_STATE);
}

/*--------------- Station state control: --------------- */

/* Station state change */
void Cm_Setstate(char S, char s)
{
    if (Cm_Getstate(S)!=s)
        st[S].reconfigure(Cm_Getstate(S),s);
    st[S].sreg=s;
}

/* Station state recall */
char Cm_Getstate(char S)
{
    return st[S].sreg;
```

```
}
/*------------- Interstation communication: --------------- */

/* Message transmission */
char Cm_Transmit(int i, char * name, char *msg) {
    port *p; char ans[2]; int sts;
    /* send the message with respect to the specified protocol */
    if (p=findPort(name)) {
        switch (p->getSyncMech())
        {
            case __BLOCKINGSEND : {
                Pearl_Sema_Try(i,p->sema_no);
                sts=p->put(msg);
                Pearl_Sema_Release(i,p->sema_no);
                break;
            }
            case __NOWAITSEND : {
                sts=p->put(msg);
                break;
            }
            case __SENDREPLY : {
                if (p->put(msg)==TRANSFER_OK) {
                    sts=p->get(ans);
                    break;
                }
                sts=TRANSMISSION_ERROR;
                break;
            }
        }
        return sts;
    }
    else
        return CONNECT_ERROR;
}

/* Message reply through the connection (part of send-reply protocol) */
char Cm_Reply(int i, char *name, char *msg)
{
    port *p; /* send the message with respect to the specified protocol */
    if (p=findPort(name))
        return p->put(msg);
    else
        return CONNECT_ERROR;
}

/* Message receipt through the connection */
char Cm_Receive(int i, char *name, char *msg) {
    port *p; char sts; char ans[2];

    /* receive the message with respect to the specified protocol */
    if (p=findPort(name)) {
        sts=p->get(msg);
```

```
        if (sts==TRANSFER_OK) {
            if (p−>sncm==__SENDREPLY) {
                ans[0]=1; ans[1]=ACK;
                return Cm_Reply(i,name,ans);
            }
        }
        return sts;
    }
    else
        return CONNECT_ERROR;
}

/*−−−−−−−−−−−−−−− System call service: −−−−−−−−−−−−−−−−−*/

void Cm_SysRequest(char S, char *sysp) {
    char msg[MAX_PARAM]; int i;
    // build msg from sysp memcpy(msg,sysp,sysp[0]+1);
    // checking the type of the station
    // if it is a TASK station, pass the system request as message to the appropriate KERNEL node
    if (st[S].type==TASK_STATION) {
        for (i=sysp[0]; i>0; i−−)
            msg[i+1]=msg[i];
            msg[0]=sysp[0];
            if (Cm_Transmit(ct[st[S].aCollection].aTask,st[S].sport,msg)==TRANSFER_OK) {
                if (Cm_Receive(ct[st[S].aCollection].aTask,st[S].sport,msg)!=TRANSFER_OK)
                    return;
                // Compose a system request message in out_param[S]
                RTOS_Cycle();
                return;
            }
    }
    // otherwise service the call locally
    if (!in_tag[0][0]) {
        in_len[0]=msg[0];
        for (int i=0; i<in_len[0]; i++) in_tag[0][i]=msg[i+1];
        RTOS_Cycle();
    }
}

void Cm_SysResult(char S) {
    int i; char msg[MAX_PARAM];
    if (st[S].type==TASK_STATION) {
        // build msg from out_tag
        Cm_Transmit(ct[st[S].aCollection].aTask,st[S].sport,msg);
    }
    if (out_tag[S][0] & 0x40) {
        // save result and continue
        tt[out_tag[S][2]].result=out_param[S];
    }
    // load context
    ct[st[S].aCollection].aTask=out_tag[S][1];
    /* initial (0) or current (1) context */
    if (out_tag[S][3]==0) tt[out_tag[S][1]].state=0;
```

```
}

void Cm_System(char S)
{
    int i;
    for (i=0; i<MAX_COLLECTION; i++)
        if (st[S].sreg==ct[i].sts) {
            st[S].aCollection=i;
            break;
        }
    if ((st[S].sreg==ct[i].sts) && (st[S].aCollection>=0)) {
        RTOS_Cycle();
        // load context
        ct[st[S].aCollection].aTask=out_tag[S][1];
        /* initial (0) or current (1) context */
        if (out_tag[S][3]==0) tt[out_tag[S][1]].state=0;
    }
}
```

E.3   Project layout

Each project is assigned a separate directory where all its specification and system configuration files are stored and maintained. The initialisation (.ini) files, comprising the system design information for the CASE environment are stored at the root of the project's directory. The data files, holding component parameters values are stored in the "\data" directory.

The listed source and header files of the system design are assembled in the project "\model" sub-directory. The "<architecture>.h" and "<architecture>.c" files reside at the root of the "\model" sub-directory. On its sub-directories, named after the individual stations of the system design, the associated station-related files reside (<Station>.c/.h, <Collection>.h/.c, <Module>.h/.c, <Task>.h/.c).

For every target platform implementation, separate project makefiles are built, which include source files from the "\model" sub-directory. Joined with the CM and RTOS libraries and compiled for the target platform, they form executable modules.

Within the Specification PEARL CASE environment, system model files are also prepared for co-simulation. They are joined with the simulator's structures accordingly, so each active unit (station, collection, task) is allotted a simulation unit in the simulation environment. The simulation traces are stored in log-files for each simulation unit separately in the "\trace" sub-directory.

The system specification in Specification PEARL syntax resides in the "\spec" sub-directory. The directory structure of a typical Specification PEARL project is laid out below:

```
<Project_name>
   |
   *.ini (files, which save the project properties and graphical design data)
   |
   <\data>
       *.db, *.px (database files, storing the properties of model components)
   |
   <\model>
       <architecture>.c
       <architecture>.h (top level architecture files)
       <\station>
           (station specific files)
   |
   <\spec>
   <Project_name>.spc (SPEARL syntax model description)
   |
   <\trace>
   *.log (log files from the co−simulation)
```

Index

© Springer International Publishing Switzerland 2016
R. Gumzej, *Engineering Safe and Secure Cyber-Physical Systems*,
Studies in Computational Intelligence 632, DOI 10.1007/978-3-319-28905-2

Printed in the United States
By Bookmasters